Cinema and Popular Geo-Politics

With a detailed range of approaches, this new collection investigates how cinematic narratives can and have been used to portray different political 'threats' and 'dangers'.

Including a range of chapters with a contemporary focus, it explores issues such as:

- how the geopolitical world has been constructed through film
- how cinema can provide explanatory narratives in periods of cultural and political anxiety, uneasiness and uncertainty.

Examining the ways in which film impacts upon popular understandings of national identity and the changing geopolitical world, the book looks at how audiences make sense of the (geo)political messages and meanings contained within a variety of films - from the US productions of Hollywood, to Palestinian, Mexican, British, and German cinematic traditions. This thought-provoking book draws on an international range of contributions to discuss and fully investigate world cinema in light of key contemporary issues.

This book was previously published as a special issue of *Geopolitics*.

Andrew Crampton teaches at Kings Norton Girls School, Birmingham, UK.

Marcus Power is a Lecturer in the Department of Geography at the University of Durham.

Cinema and Popular Geo-Politics

Edited by

Marcus Power and Andrew Crampton

Routledge
Taylor & Francis Group
LONDON AND NEW YORK

First published 2007
by Routledge
2 Park Square, Milton Park, Abingdon, Oxon, OX14 4RN

Simultaneously published in the USA and Canada
by Routledge
711 Third Ave, New York, NY 10017

Routledge is an imprint of the Taylor & Francis Group, an informa business

Typeset in Garamond by
Integra Software Services Pvt. Ltd, Pondicherry, India

British Library Cataloguing in Publication Data
A catalogue record for this book is available from the British Library

Library of Congress Cataloging in Publication Data
A catalog record for this book has been requested

ISBN10: 0–415–34818–8
ISBN13: 978–0–415–34818–8

Contents

Reel Geopolitics: Cinemato-graphing Political Space

MARCUS POWER
School of Geographical Studies, University of Bristol, Bristol, UK

ANDREW CRAMPTON
Department of International Studies, Nottingham Trent University, Nottingham, UK

> American foreign policy and film…track each other through time as compulsively as symmetrical sundials[1]

As millions of people watched the horrific spectacle of the Twin Towers collapsing after the September 11 terrorist attacks, many eyewitnesses and survivors compared their dramatic experiences to a variety of Hollywood movies. In many ways Hollywood movies provided a language and imagery that commentators drew upon in making sense of the attacks and their geopolitical implications and many Hollywood producers, directors and writers later explained that they too believed many Hollywood films had seemed to 'predict' the atrocities of 9/11. The images of destruction in films like *The Siege*, *Die Hard* and *The Peacemaker* (amongst many others) became important reference points and ways of enframing popular understandings of the radically changing geopolitical world.

In October and November 2001 Pentagon officials began a series of meetings with Hollywood directors, screenwriters, scenarists and specialists in disaster movies to 'solicit the help of Hollywood in the war against terrorism' and imagine possible future scenarios and responses to them.[2] White House advisors also met with Hollywood executives to discuss the role of Hollywood in 'getting the right ideological message across not only to Americans, but also to the Hollywood public around the globe'.[3] The subsequent release of films like *Black Hawk Down* and *Behind Enemy Lines* arguably signalled a resurgent Hollywood interest in narrating US global military adventures. Moreover, their success at the box office suggests the contemporary popular appeal of a remilitarised US global role and, in the context of the new Coalition against Terrorism, also suggest a hunger for *explanatory narratives* in this period of cultural and political anxiety, uneasiness and uncertainty. These and other films have attempted to (re)author a new post-9/11 geopolitical certainty and for some observers, these films have paved the way for further US/Coalition interventions, building support amongst the cinema-going public for a variety of global political campaigns.[4]

Of particular interest and importance in films like these are the ways in which they crystallise or firm up the uncertain role of America, American identities and US citizens in global political dramas – particularly during moments of crisis. Important questions can be raised about the ways in which geopolitical imaginations are expressed through film and about how different 'threats' and 'dangers' are constructed through various Hollywood narratives.

The articles in this special issue explore many of these themes in more depth and elaborate the complex connections between film and geopolitics. They reflect in various ways on common themes of national identity, gender and the construction of masculinity and ethnicity through film and seek to explore the importance of borders, boundaries and territories to cinematic narratives. The initial idea for a special issue emerged out of a piece of research we conducted on an earlier period of geopolitical change and uncertainty.[5] The period from the end of the Second World War through to the early 1950s saw film in the United States become a central battleground for constructing and challenging Cold War geopolitical boundaries and imaginative geographies.

During the Second World War Hollywood was 'sent to war' and produced many films such as *Days of Glory* (1943), *North Star* (1943) and *Mission to Moscow* (1943) that articulated a kind of 'Celluloid Russophoria'[6] mythologizing Russian peasant life and showing the Soviet Union as a worthy ally that was not all that different from the United States. In the changing geopolitical context of the Cold War this was reversed into a 'Russophobia' with Hollywood, under pressure and fear from the House UnAmerican Activities Committee (HUAC), producing series of 'Reds under the beds' films depicting communist infiltration in the United States. Films like *Behind the Iron Curtain* (1948), *I Married a Communist* (1949), *The Red Menace* (1949), *Big Jim McClain* (1952) and *My Son John* (1953) constructed discourses of danger that ran alongside and extended other important and official geopolitical discourses. Films of this period were intimately connected to the formation of new geographical imaginations and in this sense their imagery formed a kind of geopolitical world:

> the movies *are* a world, a country of familiar faces, a mythology made up a limited number of stories [possessing] a relation of wish, echo, transposition, displacement, inversion, compensation, reinforcement, example, warning ... What remains constant is an oblique but unbroken connection to the historical world.[7]

What also remains is a connection to the *geopolitical world*. The 1940s saw a series of geopolitical struggles and important ideological shifts and transitions as the United States reinvented itself as a benevolent defender of 'freedom' and 'democracy'. Hollywood movies offered their audiences an understanding of international politics and the role of America in this arena and brought this home to America itself, stressing the threats and dangers within and posed locally. As a medium of mass communication cinema provided an important site of competition between Hollywood, film-makers and state institutions on a 'battlefield of sounds and images'.[8] During the Cold War the discursive figuration of the 'Evil Empire' in terms of absolute difference from the United States was important in rendering and securing American identity.[9] HUAC's concern with the filmic images of America and its 'other' demonstrate that films were an important part of the attempt to secure, during the Cold War, an ideological 'fit' with a wholesome sense of Americanness, which was 100% total and complete. Following Stuart Hall's metaphor of society as a "battleground' of representations[10] we would argue that Hollywood's cinematic geographies of geopolitics became a crucial *ideological battleground* with the establishment of censorship groups like HUAC. Thus cinema was not some form of crude and unwieldy foreign policy propaganda tool but rather an important site of contestation around geopolitical meanings and scriptings. Indeed the very boundaries of national citizenship were at stake. In looking at Cold War Hollywood films our aim had been to show that Hollywood's filmic cartographies offered a certain way of reading political change and collectively amounted to a kind of celluloid fund of geopolitical knowledges. This fund of knowledges provided on-screen ammunition for sketching out the role of the US in world politics, for (re)positioning America on the global stage, and refiguring its geopolitical dramas.

Our work on HUAC and the films of the 1940s and 1950s demonstrated to us that the connections between film, identity, geopolitics and foreign policy were complex and varied but also relatively unexamined and untheorised in geographical literature.[11] Many of our sources came from film studies and historical literature – and the explicit geographical and geopolitical themes of this periods films were pushed to the margins of this otherwise rich literature. Films, we would argue, exist in an intertextual relationship with other geopolitical knowledges. They reflect, reify, explain, author, support, undermine and challenge hegemonic geopolitical discourses. In putting together this special issue we wanted to explore further the filmic authorship of geopolitics or the production of geopolitical meaning in cinematic texts and their intertextuality with other geopolitical landscapes and discourses. Our purpose has been to show that film has for a long time held a unique position in projections of identity and that these projections can provide important examples of the ways in which geopolitics is made intelligible and meaningful in the popular realm and through the 'everyday'.

A further motivation in bringing together such a volume was a concern to explore filmic narratives and cinematic traditions that existed 'outside' or beyond Hollywood and a desire to read more about the interconnections between cinema and geopolitics as they operated in other spaces and places. Previous work on colonial and post-colonial cinema in the Lusophone world[12] had highlighted the paucity of sources concerning cinema and geopolitics in Africa, Asia and Latin America and demonstrated an overwhelming focus on Hollywood as the only 'real' cinema. In many ways this special issue does contain a large number of articles which are concerned with Hollywood or Hollywood film although there are contributions here which look a little further afield – to the Phillipines, Germany and Palestine.

We propose the title 'Reel geo-politics', then, as a way of exploring the construction of a particular kind of geopolitical reasoning, a unique reading of international affairs and political or ideological differences. This attempts to build on Christensen's[13] concern with 'reel politics' in his discussion of the anti-political Hollywood films of the 1950s. We use this title not to imply that this form of the geopolitical was any more real or actual than any other but rather to signify our intention to unreel and unravel the cinemato-graphing of political space at a variety of spatial scales.

CINEMA AND POPULAR GEOPOLITICS

> As a technology of seeing and a form of projectionism film can be regarded as eminently geo-graph-ical and geo-political.[14]

Alongside the emergence in the 1990s of a post-structuralist inspired 'critical geopolitics' has been an increasing critical focus on the ways in which geopolitical knowledges saturate 'quotidian constructions of identity'.[15] Ó Tuathail and Dalby[16] usefully distinguish between 'formal', 'practical', and 'popular' geopolitics as different 'types' of geopolitical knowledges producing different but often complementary geopolitical imaginations and spatialisations of danger. The first two types refer, respectively, to the geopolitical writings of academics and the 'common sense' geopolitical statements of state leaders and bureaucrats. 'Popular' geopolitics, however, is a more expansive notion of where geopolitical knowledges are produced and refers to geopolitical representations found in a variety of cultural forms such as the popular media, novels, magazines and cartoons.[17] What is interesting about much of this recent work on popular geopolitics is that whilst novels, magazines and cartoons have received increasing amounts of attention, the medium of cinema (and to a lesser extent radio) remain relatively neglected. Yet many geographers, either in their research or in their teaching, engage film directly.[18] Cinema, as Cresswell and Dixon have recently argued, is not a traditional object of inquiry for geographers – as opposed to landscape or region – and yet, in articles and books, in lecture theatres and seminar rooms, film has become one of the most popular sites for research and teaching:

> Across the discipline, film has been used as example, as metaphor, as allegory, as a vehicle for querying the character of representation, and as a way of recording everyday perceptions of the world.[19]

Thus geographers have deployed and explored film as a 'mimetic of the real world', as a way of authenticating representations of distant peoples and places, as a way of exploring the construction of intersubjective meanings or the production of dominant ideologies and as a site of resistance. Additionally, geographers have increasingly sought to explore the complex of social processes that make up the film industry, ranging from financing to distribution and from production techniques to the construction of movie theatres and film technology. The issue of film's relationship to some exterior and pre-existent reality is central to many geographers' research agendas[20] and remains an important question in any consideration of the relationship between cinema and geopolitics, between the reel world of cinema and the real world of global political space. This involves a recognition that cinema provides an important space of confrontation and encounter for viewers and the recognition that the reception of filmic meaning is far from passive. Instead it can be argued that films raise important questions about the ways in which spaces and identities are made up and dissolved within the structure of films themselves and in the relationships between the production and consumption of space in and around film. Thus film is important in the study of critical geopolitics because it represents a constitutive element in the production of political geographies and because political spaces, places and landscapes are implicit tools in the production of film.[21] It is also necessary to recognise however that the theme of cinema and geopolitics is interdisciplinary in nature (and many of the contributions in this issue are written by non-geographers). According to Cresswell and Dixon the reality/representation divide has been gradually dissolved in recent years (particularly in the context of post-structuralist/antiessentialist interventions) leading to a reappraisal of geography's engagement with film:

> Films are no longer considered mere images or unmediated expressions of the mind, but rather the temporary embodiment of social processes that continually construct and deconstruct the world as we know it.

It is also useful here to highlight a distinction between geo-politics and geopolitics,[22] and foreign policy and Foreign Policy.[23] The hyphenation of geo-politics draws attention to the ontological unthought that makes possible geopolitics in its conventional understanding. In this sense 'official articulations of geopolitics attempt to dissimulate the geo-politics that makes them possible, to repress their conditions of (im)possibility'.[24] In a related sense Campbell's foreign policy/Foreign Policy distinction draws attention to the production of difference and otherness that takes place prior to Foreign Policy in its conventional sense. The former 'foreign policy' refers to the process of 'making foreign' and provides a discursive economy upon which conventional Foreign Policy operates – its actions reproducing and containing challenges to the identity scripted by foreign policy.

These distinctions are instructive for studies of formal, practical, and popular geopolitics but are particularly useful here because in both senses, the 'moment of geopolitics' is not fixed but diffused through society. A study of popular geopolitics recognises that 'geopolitical knowledge does not simply 'trickle down' from elite texts to popular ones' and points to the 'intertextual nature of much geopolitics'[25]and the varied sources and analogies from which popular representations of the political are drawn and come to be blurred.[26]

Geo-politics, according to Ó Tuathail,[27] is partly about 'address[ing] the big picture' and examining the ways in which actors and dramas are arranged on a world stage or a kind of 'global chessboard' of political positions. Film represents a unique way of arranging these dramas and actors and of attempting a kind of spatialisation and visualisation of boundaries and dangers and American identity is connected to the geopolitical constructions and ideological codes of Hollywood films. They provide a way of solving (geo)political uncertainty, and the very uncertain nature of America itself, through building moral geographies and making clear the lines of division between 'us' and 'them'. It is important to explore then, how in many ways cinema functions as an ideological state apparatus as the Manichean nature of geopolitical discourse was captured by the medium of cinema, and popular support for geopolitical strategies has been generated by the interpellation of movie going subjects.

THE ORGANISATION OF THE SPECIAL ISSUE

The first essay, by Elena dell'Agnese, explores cinematic representations of the US–Mexico border region over a period of 65 years and focuses in particular upon westerns, horror films and the science fiction and film-noir categories. The essay examines the portrayal of history and racial stereotypes in around 30 films concerned with the US–Mexico border region and touches upon a range of themes from a discussion of colonial fantasies to an exploration of the representation of sexuality in border films. Conceptualising many of these movies as part of a broader 'genre of imperialism', dell'Agnese deals with the cinematic representation of Mexico and Mexicans as well as the multiple ways in which borders are seen to be constituted through film (for example as 'magic curtains', spaces of differentiation/otherization, as something unsaid or unscreened or as a 'contact zone'). Discerning a number of important imaginary (political) geographies of the border and fantasies about the white American man 'going South' this essay creatively examines the way in which 'major shifts in the mood of American idealism' can be traced through these films and discusses the ways in which they might be seen as products of a particular geopolitical moment.

The next essay, by Michael J Shapiro, explores a number of classic westerns, particularly those directed by John Ford, as a way of interrogating national mythologies of the American West. Additionally, Shapiro also raises questions about a number of contemporary films which return to the same western landscapes but offer more critical engagements with both the history of Euro-American imperial expansion and the cinematic narration of the nation that characterised the earlier classic westerns. Examining the cinematic construction of Indians and Indian territory in the classic western genre, Shapiro argues that the analysis of certain films can supply insights into the Euro-American geopolitical and biopolitical 'ethnogenesis' and provide new perspectives from which to view the ethnic and territorial privileging of certain presences on the North American continent. Viewing film as an increasingly critical medium with the potential 'to challenge the hegemony of the modern state system', Shapiro explores the ways in which films about the West construct particular geopolitical spaces, reconstruct a sense of 'international relations', represent certain landscapes or reflect upon the fixity or fluidity of borders and boundaries. In detecting multiple ambiguities and an 'institutionalised silence' in many films about the West, Shapiro notes the reassertion of alternative narratives (in music, cinema and writing) and the emergence of a more visible and voluble Native American geopolitical imagination of the American continent.

The following essay, by Andrew Crampton and Marcus Power, raises questions about the role of Hollywood as geopolitician and about the importance of film as a key geopolitical site, exploring the production of subject positions through cinema. In the staging of the Second Gulf War and its multiple plots and dramas this essay argues that there was an important intertext with the Second World War and its various associations with virtue. Focusing in particular upon the figure of the geopolitician co-ordinating the production of the stage of international politics for a viewing subject, this essay explores how Hollywood narrates contemporary geopolitical space. Allied to this is a concern to explore how audiences make sense and meaning of the films they watch and the conflicts represented therein. What are the implications of drawing on filmic interpretations of the Second World War in representations of contemporary conflict? In this essay the particular example of *Saving Private Ryan* (Steven Spielberg, 1998) is discussed as a point of entry to a discussion of this intertext in an attempt to understand further how movies figure in the popular political imagination as ways of imagining and enframing global political change.

The next essay in the issue, by Klaus Dodds, focuses on the James Bond film series and sets out to explore the ways in which these films utilise particular geographical locations or seek to construct a very particular kind of geographical framing, narrating wider ideological conflicts in the process. Focusing in particular on the first five Bond films (made between 1962 and 1967) this essay examines the ways in which particular places are embedded within the filmic narratives of the Bond series in order to reflect upon the wider project of developing the conceptualisation of 'popular geopolitics'. In particular the paper considers the ways in which certain films, such as *From Russia with Love* (1963) actively created a sense of an ideological and geographical boundary between different geopolitical spaces and argues that places such as Turkey, 'were never simply passive backdrops and or "exotic locations"'. Dodds concludes that the literature on popular geopolitics could more fruitfully explore the detail of the filming process in order to understand more carefully the filmic construction of spaces of danger, threat, domesticity and exoticism.

The focus of the next essay, by Deborah Dixon and Leo Zonn is the 'perilous place of Third cinema' and the conceptual limitations within Frederic Jameson's (1992) *The Geopolitical Aesthetic*. Exploring Jameson's reading of filmmaker Kidlat Tahimik's (1976) *The Perfumed Nightmare* Dixon and Zonn adopt an anti-essentialist perspective in order to offer an alternative reading of the film and to interrogate Jameson's argument that Third Cinema is very much a redundant force within late capitalism. The reading of Tahimik's film by Jameson had prioritised resistance against capitalism and framed particular spaces within the film as meaningful and significant in regard to this project but Dixon and Zonn set out to respond to this, noting how these spaces are 'constituted from the temporary fixing of a series of people and place-based identities'. What is particularly interesting and important about this re-imagining and re-reading of the film is the exploration and recovery of geographies of gender and of how notions of masculinity and femininity are imbricated within political, economic and cultural spheres. Additionally, Dixon and Zonn also seek to focus in particular on Jameson's notion of this film as a particular instance of 'cognitive mapping' whereby the individual subject is able to experience a sense of the social totality within which they are placed. This concept is reworked by Dixon and Zonn to focus attention on the ways in which objects are embedded within a web of significations, drawing out the 'lines of fracture and contradiction that allow for such objects to become meaningful in a host of other contexts'.

In the essay by Nurith Gertz and George Khleifi, the main focal point is Palestinian cinema and its contribution (through the deployment of a variety of symbols) to consolidating the Palestinian nation and its historical narrative. Gertz and Khleifi explore the complex ways in which Palestinian cinema has attempted to create a homogenous historical narrative, to recover a lost unity and to reflect the struggle against the occupation. In particular the cinematic construction of particular territories is teased out carefully here through a concern with how different Palestinian films depict the space of the home, village or town and seek to connect this to a national or 'Palestinian space'. Tensions and contradictions between religious, familial, national and pan-Arabic identities are also explored here, particularly as they became more pronounced in films made since the end of the 1990s. Additionally, Gertz and Khleifi also concern themselves with why so many Palestinian films take place at borders and checkpoints such that they become 'road-block movies' which struggle with the complicated project of mapping Palestinian territory. Through a focus on journeys and movements between spaces and identities these films construct an image of daily life as 'stuck' and 'blocked' leading to a sense of lost space and an image of immobility and stasis. The films discussed here and the outline of the changing dynamics and landscapes of Palestinian cinema provide an important and useful window onto the escalation of Israeli–Palestinian conflict and the strengthening of the Israeli occupation in recent years.

The next essay in the issue, by Simon Philpott and David Muttimer, focuses on two 'crucial moments in the American imperium' by exploring the Americas depicted by actor Martin Sheen in his two most significant roles as the assassin Captain Benjamin Willard in *Apocalypse Now Redux* (ANR) and as President Josiah Bartlett in *The West Wing*. As Phillpott and Muttimer suggest, cinema and popular culture more generally are crucially important in the active production and the active forgetting of history and here the United States is seen to be repeatedly capable of 'remarkable acts of forgetting' and to have regularly avoided the kind of 'reflexive examination necessary for transformation in the face of an innocence lost'. The analysis of the *The West Wing* offered here explores the vision of America that is offered and the way in which it continues the self-delusion of American innocence that began with movies like ANR which focus on the loss of American innocence in the jungles of Vietnam. Interestingly, Philpott and Muttimer incorporate an analysis of the two 'crews' that surround Martin Sheen in ANR and *The West Wing* in order to reflect more widely on issues of race, gender and class in American politics. The loss in Vietnam and the subsequent questioning of the capacity for sacrifice raises some interesting and important issues about American masculinity. In particular the authors engage directly with Hollywood's struggle to reconstitute masculinity with cinematic representations of white, muscular 'hard bodies' and to depict particular ethnicities and class structures.

The next essay, by Gearóid Ó Tuathail, also raises some important questions about gender, national identity and patriotism through an exploration of John Davis's patriotic action adventure *Behind Enemy Lines*, loosely based on the story of Scott Ó Grady, a US fighter pilot shot down over Bosnia in 1995. The release of the movie in late 2001 was sped up in order to take advantage of a surge of American nationalism following the attacks of 11 September and the subsequent retaliations in Afghanistan whilst the marketing of the film relied upon 9/11 imagery. Targeted at an adolescent male audience and constructed with the PlayStation age in mind, the film stars Gene Hackman, a veteran of the US marines and several 'patriotic' Vietnam movies. In his analysis of the film, Ó Tuathail is concerned with the films invention of a heroic moral engagement and with the post-Vietnam recovery of the heroic in the context of messy historical engagements and the frustrations and confusions of the post-Cold War era. The film is explored here in order to open a window onto the contemporary culture of US militarism, to interrogate its gender politics and the historical resonance of its storyline as well as to reflect on its 'pornographic gaze on weaponry'. Additionally, Ó Tuathail is interested in the film's articulation and reproduction of the 'Jacksonian' geopolitical tradition in American geopolitical culture and also in the way the film visualises and articulates the structure of feeling of US militarism. Usefully, Ó Tuathail reminds us that movies are ultimately commercial products designed to appeal to large audiences and to make money for their producers and that as a result the complexities of geopolitics are often effaced in the name of promoting and pandering to a chauvinistic brand of militaristic nationalism.

The final essay, by Inga Scharf, explores the emergence of New German Cinema and examines the impact of post-Second World War geopolitics on constructions of West German national identity. In particular Scharf is interested in the ways in which films have represented homelessness, borders and questions of belonging and draws upon a range of recent writing about 'critical geopolitics' in order to understand these constructions and how they have been popularised and portrayed in particular movies. Raising questions about the ways in which New German Cinema seeks to articulate counter-narrations of national identity in post-war West German society and about the depiction of gender identities, Scharf argues that the analysis of New German cinema can shed light on the construction of borders and boundaries enabling us to understand more carefully the impacts of divisions and separations in particular places and within particular communities, again by bringing out the importance of journeys, exiles, crossings and exotic adventures. As in the paper by Gertz and Khleifi, Scharf is concerned with the ways in which films render these divisions, borders, boundaries and separations as something which has wide-ranging consequences for the everyday lives of citizens and communities.

NOTES

1. M.J. Strada and H.R. Troper, (1997) *Friend or Foe? Revisions in American Film and Foreign Policy* (London: Scarecrow 1997), p.xi.

2. S. Zizeck, *Welcome to the Desert of The Real*(London: Verso 2002) p.16. See also British Broadcasting Corporation (BBC), 'Hollywood: The Pentagon's New Adviser', <http://news.bbc.co.uk/1/hi/programmes/panorama/1891196.stm/>, 2002.

3. Zizeck (note 2) p.16.

4. This is hardly the first time that the United States, in response to international developments, has attempted to revitalize its propaganda activities in the Middle East. An earlier episode occurred early in the Cold War, during the Truman and Eisenhower administrations, when the United States was expanding efforts to incorporate the region into a global anti-Soviet alliance. It was concerned about the implications for US interests of the diminished post-Second World War abilities of Britain and France to project Western power and influence in the area, and by the enormous increase in anti-Western feeling that had been generated by the establishment of Israel. In Iraq, the American embassy in Baghdad in 1952 used a mailing list, mobile film units and 'certain prominent anti-Communist religious leaders' to distribute propaganda brochures. Fore more on this see Joyce Battle (ed.), *U.S. Propaganda in the Middle East – The Early Cold War Version*, National Security Archive Electronic Briefing Book No.78 (Washington DC, 2002). This article can be found at its original location, <http://www.gwu.edu/~nsarchiv/NSAEBB/NSAEBB78/essay.htm/>.

5. A. Crampton and M. Power, 'Reel Geo-politics: Hollywood HUAC and UnAmericana', unpublished paper presented at the AAG conference, New York, 2001.

6. Strada and Torper (note 1) p.43.

7. M. Wood, *America in the Movies* (New York, Dell 1975) p.3.

8. J. Hellmann, *The Kennedy Obsession: The American Myth of JFK* (New York: Columbia University Press 1977) p.97.

9. D. Campbell, *Writing Security: US Foreign Policy and the Politics of Identity* (Manchester: Manchester University Press 1992).

10. S. Hall, 'Notes on Deconstructing "The Popular"' in R. Samuel (ed.), *Peoples History and Socialist Theory* (London, Routledge 1981), pp.227–40; idem, 'Cultural Identity and Diaspora', in J. Rutherford (ed.), *Identity: Community, Culture, Difference* (London: Lawrence and Wishart 1990), pp.222–37.

11. There have been some important exceptions here, notably Frederic Jameson, *The Geopolitical Aesthetic: Cinema and Space in the World System* (London, BFI 1992) and Eric Cazdyn, *The Flash of Capital: Film and Geopolitics in Japan* (Durham, NC: Duke University Press 2002); see also J. Sharp, 'Reel Geographies of the New World Order', in G. O Tuathail and S. Dalby (eds), *Rethinking Geopolitics* (London: Routledge 1996).

12. M. Power, 'Geo-politics and the Representation of Portugal's African Colonial Wars: Examining the Limits of 'Vietnam Syndrome', *Political Geography* 20 (2001) pp.461–91; idem., 'Post-colonial Cinema and the Reconfiguration of *Moçambicanidade'*, *Lusotopie* (forthcoming 2005).

13. T. Christensen, *Reel Politics: American Movies from Birth of a Nation to Platoon* (London: Blackwell 1987).

14. G. Ó Tuathail, '(Dis)placing Geopolitics: Writing on the Maps of Global Politics', *Environment and Planning D: Society and Space* 12 (1994) p.540.

15. G. Ó Tuathail and S. Dalby, 'Introduction: Rethinking Geopolitics: Towards a Critical Geopolitics' in Ó Tuathail and Dalby, *Rethinking Geopolitics* (note 11) p.5.

16. Ibid.

17. Ibid.; K. Dodds, *Geopolitics in a Changing World* (Harlow: Pearson Education 2002).

18. T. Cresswell and D. Dixon, 'Introduction: Engaging Film', in T. Cresswell and D. Dixon (eds), *Engaging Film: Geographies of Mobility and Identity* (London: Rowman and Littlefield 2002).

19. Ibid. p.1.

20. Ibid.

21. Ibid.

22. G. Ó Tuathail, '(Dis)placing Geopolitics: Writing on the Maps of Global Politics', *Environment and Planning D: Society and Space* 12 (1994) pp.525–46. G. Ó Tuathail, *Critical Geopolitics: The Politics of Writing Global Space* (London: Routledge 1996).

23. D. Campbell, *Writing Security: US Foreign Policy and the Politics of Identity* (Manchester: Manchester University Press 1992.

24. Ó Tuathail, *Critical Geopolitics* (note 22) p.67.

25. J.E. Sharp, *Condensing the Cold war: Reader's Digest and American Identity* (Minneapolis, MN: University of Minnesota Press 2002), p.35.

26. J. Der Derian, *Virtuous War: The Military Industrial Media Entertainment Network* (Boulder, CO: Westview Press 2001); idem, 'Global Events, National Security and Virtual Theory', *Journal of International Studies* (Millennium), 30/3 (2001) pp.669–90; M. Shapiro, 'The Sports/War Intertext', in J. Der Derian and M. Shapiro (eds), *International/Intertextual Relations* (New York: Lexington 1989).

27. G. Ó Tuathail and S. Dalby, 'Introduction: Rethinking Geopolitics: Towards a Critical Geopolitics;' in G. Ó Tuathail and S. Dalby (eds), *Rethinking Geopolitics* (London: Routledge) p.1.

The US–Mexico Border in American Movies: A Political Geography Perspective

ELENA DELL'AGNESE
Department of Sociology and Social Research, University of Milano-Biococca, Italy

INTRODUCTION

Along with the open spaces of the West and the fragmented spaces of New York and Los Angeles, the region surrounding the US–Mexico divide is probably one of the most frequently screened landscapes of North America. Along with them, it represents a way of 'narrating' the nation,[1] that is, of embodying its features and qualities in a certain place, or image of place. However, the cinematic versions of Monument Valley and the North American cityscape usually promote (or challenge) the many dimensions of the US national narrative,[2] while the cinematic border – with its function of interfacing (what is considered to be) the Self with (what is considered to be) the Other – figures as a prominent feature in the making of a more internationally oriented geopolitical discourse, narrating the United States in its relations with the Outside.

Certainly, 'the border was a common setting since the beginning of commercial cinema'[3] and American–Mexican relations have been a major theme since the foundation of the film industry.[4] Indeed, the Mexican Revolution of 1910–20, with the dramatic events of the raid led by Pancho Villa on Columbus (New Mexico) in 1916, and of the ensuing punitive expedition, which in the same year saw Villa pursued nearly 300 miles south of the border, coincided with the rise of feature cinema as the dominant form of visual entertainment in the United States.[5] At the time, movies not only provided the American public with powerful images of the region,[6] but also produced a popular discourse about the geopolitical reasons for the events. 'In 1916 alone, seven feature films were released that explicitly dealt with military mobilisation on the border.'[7] They were quite racist productions in which Mexicans were represented as 'greasers' and amply vilified,[8] while the Americans were generally shown as heroes.

Since then, the border has been a leading or secondary character in dozens of movies, ranging from the obvious westerns to the less obvious horror, science fiction and film-noir categories.[9] As with any other 'reel landscape',[10] the US–Mexico border has been invented and re-invented by the camera, which has selected certain clichés and developed its own topoi. Through the century, the border was depicted as a racialised space, as a gendered area where masculine and feminine stereotypes can be portrayed against a backdrop of colonial fantasies, as a place of contested historical heritage and even as a symbolic landscape of cultural hybridisation and encounter. Because of these many race and gender implications, its narratives and imagined geographies have been the object of research in a wide range of academic fields, from American and Chicana/o studies to literature, film, gender and cultural studies, and cultural geography.

However, the border is, above all, a politically charged place, and its portrayal in popular culture also deserves to be analysed with the instruments of political geography. Indeed, the same signifier (the 'border') can have different meanings: in terms of political geography, it implies both the boundary line demarcating the end of sovereignty and the zone of interchange between one state and the other. It can also mean a barrier limiting the free movement of people and goods, a symbolic end of the nation-Self and/or an open frontier.[11] A different line of reasoning is involved in each of these different meanings.

In this contribution, I will try to highlight the connotations that the American film industry has given to the cross-border experience over the years – both from the south and north – and the meaning of the border that has been assumed to support such interpretations.[12] From a 'critical geopolitics' perspective, I will assume that such connotations are popular forms of expression of a geopolitical discourse,[13] pertaining to the role and standing of the United States relative to its southern neighbour.

In order to plumb the deep structure of this geopolitical discourse, I will try to identify the framework upon which it has been built, considering a miscellany of about 30 movies, all produced in North America over a period of 65 years. All these films have to do with the border, the border region or the crossing-the-border experience, and may be arranged into three or four different groups: the 'south of the border' adventure, illegal immigration and trafficking, and the question of cultural separatedness, hybridisation and encounter. Some of these movies are very famous – works of art by celebrated directors, such as Anthony Mann, Orson Wells and John Huston – and have been discussed elsewhere by reviewers and academics. Others are Hollywood blockbusters that have received greater attention from audiences than from critics and intellectuals. Others are merely B movies, that is, low budget productions with little merit and fewer ambitions. Together they form a remarkable body of work belonging to what is commonly defined as 'popular culture' – that is, to the everyday practices, beliefs and 'geopolitical imagination' of 'common people'.[14]

SCREENING THE FRAMES I: GOING 'SOUTH OF THE BORDER'

South of the Border (1939, dir. Sherman) is a classic pre-war western starring Gene Autry, 'America's favourite singing cowboy'. It can be taken as a starting point for analysing cinematic representations of the 'crossing the border' experience from a North American viewpoint, since it variously assembles all the ingredients later to become the underpinnings of a long series of movies presenting similar views. It belongs to a generation of movies built around a confusion between the mythic space of the western and the real issues of the day.[15] In terms of popular geopolitics, such movies produced a powerful interaction between icons of the past and present, thereby suggesting that 'the heroic ethic of Western vigilantism had a kind of timeless validity as a means of resolving a social or political impasse'.[16] *South of the Border* is, from this perspective, absolutely intriguing. It is constructed around the character of Gene, who, even if dressed as a cowboy and riding a horse, is a special federal agent on the eve of the Second World War. At the beginning of the story, the singing cowboy is riding south of the border with his pals against open scenery of vast rugged lands and cactuses. He arrives in a small village, where he enjoys the local atmosphere of fiestas and fun, and starts romancing a local *señorita*. But then, following government orders, he is sent farther south to the fictitious region of Palermo (Mexico) to foil the plans of foreign spies who are attempting to gain control of the local oil fields. At the end of the film, he goes back to his *señorita*, only to discover that she has become a Catholic nun. He therefore rides back north with his pals.

In synthesis, the film suggests that for the white American man, going south of the border: (1) is always an option, (2) will guarantee an exotic experience and maybe a chance for adventure (and if the situation gets too dangerous, there is always a safe way back), (3) will provide easy romance/sex with local girls, (4) will offer a chance to exert American moral superiority and technological ability in a less developed and civilised context, and (5) will represent a 'rite of passage' to adulthood for the individual, as well as for the nation.

All these elements are present, even if variously intertwined, in most of the movies about south-of-the-border adventures, notwithstanding the many different cinematic genres and subgenres they may belong to. However, thrillers, horror movies and sex comedies are usually more tuned to sex and travel, while westerns tend, in the main, to carry a more or less explicit 'political' discourse.[17]

Indeed, the Rio Grande/Rio Bravo region between Texas and the Mexican border is a common setting for westerns and is repeatedly depicted as an open landscape of vast wilderness in order to emphasise the regenerating role of the frontier experience.[18] Since *The Caballero Way*, an early silent film about the Cisco Kid,[19] westerns set in Mexico have become sufficiently numerous to represent a specific subgenre in their own right. They include such major successes as *Vera Cruz* (1954, dir. Aldrich), *The Magnificent Seven* (1960, dir. Sturges) and *The Wild Bunch* (1969, dir. Peckinpah). Many of these Mexico westerns are so-called 'professional plot westerns'[20] and centre on the adventures of professional fighters taking on an assignment south of the border for money. Major shifts in the mood of American idealism – and in the representation of the nation's geopolitical mission in the rest of the world – can be discerned from one movie to the next: while *The Magnificent Seven* can be described as the narrative of an altruistic crusade, *The Professionals*(1966, dir. Brooks) and *The Wild Bunch* are infused with cool cynicism mixed with nostalgic eulogy to the past. Nevertheless, all these professional movies share 'a similar albeit selective code of honour'.[21] So, notwithstanding the mercenary attitude displayed at the beginning of the narrative, while in Mexico, the American 'professionals' cannot avoid becoming involved in the fight of the indigenous people for liberty and freedom, and end up fighting against the local tyrant for the sake of justice and not money.[22] Even *The Wild Bunch*, apparently misanthropic and anti-American in its message, does not break the formula: the members of the Bunch may be depicted as brutal bandits at the beginning of the story, but, through their Mexican adventure, they undergo a regeneration (symbolised by Peckinpah's use of landscape imagery and juxtaposition of confined and open spaces). Indeed, they rise from their 'lowlife selves to the level of heroes',[23] who fight their last battle to advance the cause of the revolution.[24]

In more recent Mexican westerns, elegiac tones tend to prevail over overt political messages. Nevertheless, stereotypes about the cross-border experience are no less abundant. *All the Pretty Horses*, shot in 2000 and starring big names such as Matt Demon and Penelope Cruz, is a good example. The narrative is a loose adaptation of the first novel of Coran McCarthy's 'border trilogy'.[25] It relates the coming-of-age of two young cowboys who, after their ranch in Texas is turned into an oilfield, ride to Mexico in search of the kind of rural society that is disappearing at home. Notwithstanding the beauty of the landscape, the two discover a much less Arcadian reality than expected, finding a ruthless society, where the strongest is always right and justice is biased. After many adventures, including confinement in an extraordinarily violent jail and the cold-blooded shooting of another teenager by the Mexican police, they manage to go back home, but not without a tortured romance between the Latin beauty Alejandra (Penelope Cruz) and the white American Jon Grady Cole (Matt Damon).

The love and sex theme appears to be the staple of many other productions about the south-of-the-border experience. Indeed, very few crossing-the-border movies refrain from mentioning the opportunity of sex and romance. Even the cartoon *The Three Caballeros* (1945) offers a Disneyfied version of the land south of the border as a place for hedonistic revelry and sexual abandon, with Donald Duck shouting 'Hot stuff!' when presented with the vista of Mexican beauties at the beach. A similar perspective is presented in the more recent *The Mexican* (2001, dir. Verbinsky), in which the main character, Jerry (Brad Pitt), is invited by his boss to go to Mexico to look for an antique gun, asking him, 'Do you like sex and travel?'.[26]

Some movies specifically centre on the sexual aspect of the Mexican adventure (leaving pornographic productions to one side, of course). A good example is *Losin' It* (1983, dir. Hason), a sex comedy that presents the classic Mexican western adventure. The film is the story of a shy young man (Tom Cruise) who decides to cross the border with a group of friends in a quest to lose his virginity. Instead of riding a horse, they are driving a beautiful convertible. And instead of searching for open spaces, they are looking for the closed landscape of the city. They nevertheless enjoy the usual type of adventures. They also spend time in a horrid Mexican jail and have to face many complicated circumstances that blatantly depict the border-crossing experience as a rite of passage to masculinity.

Another movie of interest in this vein is *From Dusk till Dawn* (1996, dir. Rodrigues). It is not a sex comedy but a sort of splatter/pulp movie, in which the border is a symbol of the transition to adult life and the passage from life to death (and eventual resurrection). Its narrative and structure are compound, with the movie constructed as a splatter, Tarantino-style film in the first 'American' half and as a horror movie in the second 'Mexican' half. The story narrates the escape of two vicious criminals, the Gecko brothers (Seth, played by George Clooney, and Ritchie, played by Quentin Tarantino), who leave a trail of corpses behind them on their way to the Mexican border. Over the border, the criminals turn victims as they stumble across a biker bar called Titty Twisters, where beautiful Mexican women dance naked for the pleasure of the male clientele and that proves to be nothing less than a house of vampires. By fighting the hideous creatures – complete with frenzied images of disembowelled and decapitated female bodies[27] – the two killers undergo a baptism of fire and purification through blood. While the weaker of the two, Ritchie, is unable to resist and becomes a vampire himself (and in this new role has to be destroyed by his own brother), the most determined of the pair, Seth, is victorious. At the end of the night, as a kind of new Cortés, he exterminates the monsters, recasting himself from villain to hero. Symbolically dead as a killer, he is reborn as a man. The image of the border as a 'magic curtain' through which one passes into a 'different world'[28] is pushed to the extreme, it marking a change in life itself.

SCREENING THE FRAMES II: FROM SOUTH TO NORTH

Whereas all the films discussed to this point present crossing the border southward as an easy option, attempts to go *El Norte* from the other side are usually shown as being much more difficult and requiring getting through a much less penetrable barrier. Mexican narratives are more complex and tend to treat the border as a limit as well as a gateway.[29] American narratives, on the other hand, usually restrict their descriptions of the 'south to north experience' to illegal immigration (and, as we will see later, drug trafficking).

Probably the first commercial movie about this sensitive topic was shot in Mexico. It was *Pito Pérez se va de bracero* (1947, dir. Riera), a comedy-drama in which the villains are a group of Italian 'coyotes' (*coyote* being the term used for immigrant smugglers). However, the American film industry tackled the issue during the same period with a more political agenda. The first relevant Hollywood film in this vein was *Border Incident* (1949, dir. Mann), a film noir with an abnormally large amount of violence for the time. The film shows the illegal traffic of *braceros*, their brutal disposal by the very men who were supposed to recruit and protect them, and the (successful) attempts of the Federales and the American forces to stop the traffic. Notwithstanding its attempt at being anti-racist,[30] the movie typecasts Mexico with the usual stereotypes. Thus the technology of the American police is constantly compared with the Stone Age primitivism of the Federales, an attitude exhibited in other movies from the same years. For instance, in *The Treasure of the Sierra Madre* (1948, dir. Houston), one of the characters says, 'The Federales don't operate in our American way. They're not fingerprint experts, but they can follow any trail'. And the Mexicans are often pigeonholed as the classic *estúpido* (again, we can compare *Border Incident* to *The Treasure*, where one of the characters says, 'Americans may be greedy, but the Mexicans are so stupid they can't tell gold from sand').

Border Incident was the benchmark for a long series of American movies dealing with crossing the border from the Mexican side. Since its release, Hollywood productions have shown going to *El Norte* as a question of legality/illegality, immigration, smuggling and law-enforcement – a set of issues never tackled when moving in the opposite direction. While illegal immigration is an issue that 'open minded' film-makers dramatise for the American public so as to create a benevolent attitude to the poor immigrants, their good intentions are invariably paternalistic towards the 'poor peasants'. Moreover, in all these movies the American male is always central, their main characters being invariably white or, as we will see in more recent productions, white-looking. They always handle the situation from the US side and always fall in love – openly or secretly – with a female illegal who needs their protection. Plus, the difficult circumstances faced by the illegals, their quasi-enslavement, the risks they run and the brutality of their treatment never fail to suggest a clear 'stay at home' discourse. Also as regards the landscape, the scenery of untouched nature and beautiful open spaces of the south-of-the-border movies is turned into a fragmented backdrop, visually dominated by physical barriers and divides (images of the border range from the river to a barbed wire fence and a daunting patrolled wall).

The list of movies on this subject is quite long and includes expensive productions starring big Hollywood names, such as *The Border* (1980, dir.

Richardson) with Jack Nicholson, as well as B movies and independent productions. Sharing a general moralistic stance, these movies, when seen chronologically, portray an evolution in the treatment of the figure of the coyote. Cast as the unqualified villain in early productions, the coyote has progressively transformed into an almost moral figure and, more recently, the ultimate western hero. In *Border Incident*, for example, *braceros* are presented as the victims of the tragic situation created exclusively by the greed of the smugglers. Indeed, notwithstanding their role in making the border so difficult to cross, the United States and Mexico are shown cooperating to stop the traffic, whose only criminals are shown as the brutal 'slave merchants' (as the movie title was aptly translated for the Italian version). In *The Border*, Mexicans continue to be shown as victims, but the relationships among patrolmen, coyotes and Mexicans are slightly more multifaceted. American border patrol agents, for instance, can be corrupted and they collaborate with the coyotes (such as Jack Nicholson's partner, who asks for money in exchange for letting immigrants in). At the same time, the coyotes are represented as high-minded characters, smuggling illegal immigrants into the States but not hesitating to shoot anyone carrying drugs.

An up-to-date version of the coyote is presented in the B movie *On the Borderline* (2000, dir. Oblowitz), which combines the two main 'border dramas' of our time: people smuggling and western nostalgia. It is the story of a young married couple, Nicky and Luke, from Detroit, who, on the drive to Pomona, California, stop at the border town of El Corredor (marked by street signs as 'The Edge of America'). Their money is stolen, so to earn some dollars to reach their final destination, Nicky starts working as a waitress in the town's diner, while Luke, unable to find anything better, agrees to help the local coyote smuggle immigrants. Despite being forced into this situation, Luke remains one of the 'good' guys of the story, along with Nicky and Connie, a motherly Mexicana. In the tradition of the genre, the coyotes could be expected to play the villains. And villains they are. Nevertheless, the most captivating character is Dean, the head coyote, who is portrayed as a dashingly handsome young man (by actor Bill Sage), able to steal the couple's money, almost seduce Vicky simply by dancing with her, force Luke to participate in the smuggling, kidnap the girl, and almost kill her. He is shot at the end, but not without reciting an elegiac hymn to the West in what is probably the most interesting sequence of the movie: 'People think the West is dead. It just doesn't look like a John Wayne movie anymore. It ain't dead. It's just updated. We traded our horses in for pickup trucks.' 'And your cattle for illegal immigrants?' 'That's right.'

More conventional in its characterisation of the coyotes as the bad guys, but innovative in the casting of the (male) hero, is *The Gatekeeper* (2001, dir. Frey), an independent production, written, directed and acted by former University of San Diego student John Carlos Frey. The movie, which won nearly ten awards at various film festivals around the United States, takes its title from Operation Gatekeeper, a program introduced by the US government in the early nineties aimed at limiting undocumented workers from crossing the border. It is the story of Adam Field (played by Frey), a half-caste US patrol agent who, thanks to his good looks and green eyes, succeeds in posing as an American, hiding his half-Mexican origins even from his American fiancée. In his attempt to become perfectly white, he gets involved in a white racist organisation that acts against immigration. Following their suggestion, but without any support from his superiors, Adam goes undercover on the other side of the border to expose the illegals' routes into the States. His plan, however, goes awry and instead of being rescued by the organisation as he is crossing the border, he ends up, together with other immigrants, in the frightening reality of a sort of enslavement camp in central California. Here is forced to work in a clandestine laboratory producing drugs and also witnesses the rape and killing of a Mexican woman by a gang of coyotes. He obviously cannot avoid taking the side of his fellow victims. The movie poses interesting questions about the quest for a white identity and the self-imposition of white supremacy beliefs by a half-Mexican character; but it cannot help falling into the usual mould, screening Mexican women as the inevitable victims, Mexican men as basically unable to protect them and the (white-looking) American hero as the only character strong enough to deal with the situation.

Cross-border immigration as a cinematic subject is not only reserved to pseudo-documentaries and dramas. Other Hollywood productions have variously referred to the subject. Two science fiction blockbusters deserve mention: *Men in Black* (1997, dir. Sonnenfeld) and *The Day after Tomorrow* (2004, dir. Emmerich). Science fiction, for the chances it offers to depict a distorted – and sometimes reversed – portrait of its contemporary reality, is a very suitable vehicle for addressing sensitive geopolitical issues and discourse.[31] *Men in Black*, which is set on a planet Earth populated by human beings and aliens in different disguises, is a clear metaphor for the US immigration question. The issue is explicitly tackled in the opening scene, in which the border patrol stops a truckload of illegal immigrants being smuggled across the border into the United States. The movie resolves the issue of immigration in terms of assimilation and blending (to be accepted, 'good aliens' are supposed to be properly disguised as humans, preferably white, while bad aliens are constantly metaphorically referred to as 'bugs').

The Day after Tomorrow is another highly political movie, aimed at the environment debate and US environmental policies. Set in the present day, the movie depicts the sudden and catastrophic onset of a new Ice Age, supposedly caused by global climate changes that will follow the greenhouse effect. It therefore openly attacks the US government's decision not to sign the Kyoto agreement and, more in general, accuses the developed world of being the cause of global warming and the resulting global freeze. Indeed, the developed world is also the north of the planet. It is shown as the principal victim of its own inconsiderate environmental policies: in the new Ice Age, the north will be entirely covered with ice, while the south will have a more temperate climate. In this situation, the only chance for survival for the American people will be to flee across the border into Mexico. The movie therefore shows thousands of refugees trying to cross a suddenly hostile border – and Mexican authorities relenting to the tidal wave of immigration only after the US promise to write off all Latin American debts, as if to say that the freedom of Americans to cross the border is a matter of luck and not a given.

SCREENING THE FRAMES III: TRAFFICKING (FROM SOUTH TO NORTH, AGAIN)

The US–Mexico border not only represents a barrier to human mobility (at least from the southern perspective), but it is also a marker of sovereignty, denoting the different systems of laws and values of the two states. Flows of goods are controlled, too, in some cases being restricted by various regulations and taxes and, in others, supported by varying degrees of strictness in the application of similar rules. At any rate, the smuggling of illegal goods is quite a common activity. Since the first half of the last century, one of the main problems of trans-border relations has been the import of drugs into the United States from Mexico. Hollywood has not missed the chance of focusing on this theme.

Of the long series of films centred on cross-border drug traffic, one of the first is probably *Borderline* (1950, dir. Seiter). It is the story of two undercover agents (Fred MacMurray and Claire Trevor) who infiltrate a drug smuggling ring in Mexico. But since neither of them is aware of the other's identity, they suspect each other of being crooks. Although spiced up by the obligatory love story between the two main characters and some sparkles of humour, the movie is quite racist. The Mexicans are constantly depicted as *estúpido* and sleepy or, if awake, as distracted by some triviality rather than doing their job properly. The Mexican police, for example, are very easy to set up, while all other Mexicans are cast as servants or drivers – with the females longing for white males. Mexicans are not even capable of being villains, with the head of the gang an American character (played by Raymond Burr).

Although not the main theme, drugs also play a significant role in Orson Wells' *Touch of Evil* (1958). The movie, set in a squalid border town, centres on the interaction between two policemen: Mexican investigator Mike Vargas (Charlton Heston) and American police chief Hank Quinlan (Orson Welles). Vargas, a Mexico City policeman, is in town with his American wife, Susan (Janet Leigh), to investigate the local traffic in drugs and arrests a member of the drug-dealing Grandi family. Meanwhile, a murder, organised on the American side of the city, is committed on the Mexican side. The main suspect appears to be Sanchez, a young Mexican who has a relationship with the daughter of the victim, a rich American businessman. But Vargas suspects Quinlan of planting evidence against him and begins investigating the police chief. Meanwhile, the Grandi family intimidates Vargas by kidnapping his wife, who is threatened with rape and set up as a drug addict. The movie, underrated at the time of its release but today regarded as a masterpiece, has many attractions besides its quality and style. In addition to Quinlan's legendary line, 'Let's go back to civilisation', which he says when leaving the Mexican side of town, and Vargas' famous, 'This is not the real Mexico ... Border towns bring out the worst in a country', the movie offers a highly symbolic cityscape – as fragmented and as complex as the relationships in the city – and more than one stereotype. From the beginning, it suggests that Mexico is a land of easy sex, opening with a shot of a neon sign advertising 'Girls'. The entire episode of Susan's kidnapping hinges on the presumption that Mexicans are sexually promiscuous and prone to the abuse of illicit drugs. Even the comparison between the two policemen makes the American look the better character. Quinlan, although corrupt and grotesquely fat, is indeed the strong man of the story, while his Mexican counterpart, notwithstanding his good manners and better intentions, not only fails as an investigator, taking the side of a culprit, but also as a husband, his inability to protect his wife representing his 'ineffectual masculinity'.[32] However, the movie at least recognises that Mexicans are capable of being masters of their own lives. They are not typecast simply as extras or servants, as in *Borderline*, but are equal to their role as villains.

More then 40 years later, a similarly complex effort to deal with the subject was attempted by Steven Soderbergh in *Traffic* (2000), an all-star Hollywood blockbuster that won four Academy Awards (including best director, and supporting actor for Benicio Del Toro). Shot in many different locations, sprawling from the slums of Mexico to the corridors of power of Washington DC, *Traffic* tells four different stories, all interconnected by the common theme of drugs. Although the many different locations were chosen to convey the pervasiveness of drug culture and trafficking in America, the movie begins at the Mexican border and develops three of its narratives in the area between Tijuana and San Diego. Even in the fourth story, set in Washington, Robert Wakefield, the head of the American drug force, played by Michael Douglas, visits the border.

The first scene of the movie introduces Javier (Benicio Del Toro) and his partner Manolo (Jacob Vargas), two moderately honest Mexican cops, as they stage a drug bust near the border. It is immediately made clear, however, that they lack the means to pull off the bust: first, the smugglers tell each other, 'Don't worry, it's just the police', then a Mexican Army squad takes over the arrest – later it is discovered that the head of the Mexican Army, General Salazar (Tomas Milian), is working against the Tijuana drug cartel on behalf of another drug cartel from Mexico City. Other interconnected narratives centre on Robert Wakefield's teenage daughter, who, despite her father's job as head of the drug force, is a drug addict herself. On the flip side, there is drug lord Carlos Ayala's glamorous wife Helena (a pregnant Catherine Zeta-Jones), who is oblivious of the source of her husband's wealth until he is arrested, but then quickly takes over the business in order to maintain her lifestyle. Meanwhile, two US cops (Luis Guzman and Don Cheadle) are handling Eduardo (Miguel Ferrer), a mid-level trafficker in Carlos' organisation who promises to be a fundamental witness in the case.

In *Traffic*, the intricacies of the plot are accompanied by many subtleties, making the movie quite elegant in its portrayal of a large number of characters and multiple story lines. The landscape becomes a symbol through the use of different coloured filters: images of the Mexican landscape are coloured with a washed-out yellow, while the North American scenes are shaded with a cold blue. Only San Diego is represented in real life bright colours, as if to imply that the borderscape reality exists as the coming together of the two complementary sides. The movie is also loaded with clear political messages: in the face of such a pervasive and devastating social problem as drug trafficking and use, Washington politicians are 'a cadre of Gucci-loafered hypocrites', as remarked by an enthusiastic reviewer, while law enforcement is bound to fail. The whole idea of curbing drug traffic by enforcing the law seems questionable. Indeed, the only results the police get are from information they get not from good-hearted civilians, but from competing drug cartels. The amount of money made by the cartels gives them plenty of resources and the police cannot compete. Soderbergh's warning is, therefore, not to combat drugs by prohibition but through preventive measures and caring for young people who need it. Notwithstanding its good ideas, the movie is not without its racial stereotypes. In fact, all the Mexican policemen, apart from Javier and Manolo, are portrayed as corrupt and brutal. By contrast, the American police, while inefficient, are invariably honest. In addition, drug dealers are shown as black people living in inner cities, while white people are cast either as distracted spectators of the tragedy (such as the Washington politicians) or as victims (such as the judge's innocent and pure daughter). None of them is involved in the traffic (the California ring is in the hands of a character named Carlos). This is the opposite of *Borderline*, proving that racism may change its mask but never dies.

SCREENING THE FRAMES IV: IN-BETWEENNESS

Of course, depicting the border not only involves crossing it. In film, *la frontera*is also a place to live, a region where you can work and sometimes make money. The borderland atmosphere is powerfully captured in movies such as *The Treasure of the Sierra Madre*, *Touch of Evil*and *Traffic*, which all depict a unique place of opportunity that also thrives on vice and drugs.

The border not only marks the limit of sovereignty. It is also a national marker that, after being superimposed across a landscape that is identical on either side, has become so entrenched in the local cultural landscape as to become of cultural significance in itself. 'The border is a boundary line between nations that are meant to be different'.[33] Therefore, living on the border means living in the midst of a process of differentiation – that is, in the place where opposites are produced so as to distinguish one side from the other (masculine/feminine, civilisation/barbarism, master/servant and so on).[34] Simultaneously, the border is a place of encounter and a paradigm for intercultural exchange and transcultural crossing and mixing. Living on the border means experiencing hybridisation in a region that may have different visual forms and norms but cannot avoid a state of in-betweenness.

The border as a contact zone has long been disregarded in film. Manufacturing differences has been the most common option for American filmmakers, while typecasting border figures as 'the trope of absolute alterity' has made the border a landscape of 'otherisation'.[35] Only in the nineties has the question of hybridisation and multiple identities become a subject of interest, the major work in this vein being John Sayles' much acclaimed *Lone Star* (1996). An independent production, *Lone Star* went almost totally unnoticed in Italy and many other European markets, but has been much acclaimed by American reviewers and audiences. The movie is set in the (fictional) town of Frontera, a Texas border town mostly populated by Latinos. When some skeletal remains are discovered in the desert, Sheriff Sam Deeds (Chris Cooper) investigates. He believes his own father, the town's previous sheriff, may have been the murderer, since the bones are those of Charley Wade, the corrupt sheriff his father was reputed to have run out of town. But his father, the archetypal hero of the 'white West', was a legend in his time, and now the city council, under the illusion of re-establishing white supremacy in the city, plans to name the new courthouse after him – despite the fact that Sam is bound to be the last American sheriff in town. Various political, racial and family tensions in the small community are revealed through the experiences of other African American and Mexican characters. The most striking of these is Pilar (Elizabeth Pena), the Mexican-American teacher who was Sam's girlfriend and is now facing a school board angry about her multicultural approach to the history of the Alamo. *Lone Star* has been quite widely discussed by academics[36]and certainly warrants it for its progressive stance, its depiction of the border region in all its ethnographic complexity, and, more importantly, the attention it devotes to such fundamental issues as received history. The movie is constructed around the themes of the shaping of memory (and memories) and the making of place(s) – symbolic landscapes becoming not only a feature of the film but also the main theme of the narrative. Despite its many merits, the movie has received criticism as well.[37] Rosa Linda Fregoso, for example, remarks that whiteness is still privileged, and, although it is 'no longer the white racist masculinity that framed race relations in a previous era, a new benevolent patron *amigo* of Mexicans and blacks is figured in the personas of Buddy and Sam Deeds'.[38] The film may tell the audience that the time of the arrogant white character is over, but being white still seems to be an advantage. In addition, the ethno-racial encounter is represented allegorically through the reunion of the two lovers, Sam and Pilar, and this is still a white man–brown woman relationship, which, in Fregoso's terms, can be regarded as a way of recycling 'colonialist fantasies'.

IMAGINARY (POLITICAL) GEOGRAPHIES OF THE BORDER

As a contested landscape and a multidimensional place, the border between the United States and Mexico has been presented and re-presented by the American film industry in many different guises and with many different interpretations. These films have ranged from the openly racist movies of the period of the Columbus raid and the punitive expedition, in which Mexicans had to be portrayed as inferior enemies, to the melodramas of the post-Second World War period, the political 'Mexico westerns' of the 1960s and the progressive movies of today. They have moved forwards from making Mexicans look inferior to examining acquired meanings and otherising processes. Notwithstanding, stereotypes about gender and race appear to be very difficult to eradicate. American movies, even when presenting a progressive political discourse, seldom succeed in escaping from standardised categories of inferiority/superiority. So, when confronting – on any level – the question of international/inter-ethnic relations, they almost inevitably present a white male, possibly of American descent, in the leading role.

The North American white supremacy discourse also figures in the cinematic representation of the border. Indeed, the US–Mexico divide can be portrayed in numerous different ways that fall, from the perspective of political geography, within different definitions of the notion of *border*. In a movie such as *Lone Star*, the border is represented as a borderland (a *trans-frontera*) – that is, as a region that is culturally unique and the by-product of the meeting of two different cultural regions. It is a place for encounter and transculturation, which can be defined, using Mary Louise Pratt's terminology, as a 'contact zone',[39] in which *mestizaje* and cultural syncretism are the norm rather than the exception. However, even in a situation of contested identities, undermining the privileged position of whiteness as the medium for interracial contact seems quite difficult to achieve: it is still only the white man who can cross the racialised border.

Movies such as *Touch of Evil* and *Traffic* treat the border as a region too, but, instead of being focused on questions of race, gender and culture, they look at cross-border relationships based on economic transactions. The border as a marker of different sovereignties allows – and stimulates – illegal trafficking, smuggling of forbidden goods and lawlessness.[40] The borderland presented in these movies is therefore a zone whose uniqueness lies in the economics of crime and vice. In such as scenario, which certainly 'brings out the worst' in the two countries, Mexicans may or may not be cast in the role of villains, but they are invariably relegated to a subordinate position, both in terms of race and gender: if women, they are the marketable objects of easy sex; if men, they may be variously typecast as corrupt, stupid or ineffectual, but never the hero of the story.

When the question of the movement of migrants from Mexico into the United States is tackled, the border tends to lose its regional dimension and is depicted as a political divide. In terms of political geography, movies such as *Border Incident* (1949), *The Border* (1980) and *The Gatekeeper* (2001) represent the border as a demarcated boundary – that is, a line that embodies the political separateness of the two states. No longer a zone of encounter or trafficking, the border acquires the characteristics of a barrier. Established to delimit the territory of the United States, it can be crossed by non-US citizens only with the (patronising) permission of the Americans. From this perspective, illegally crossing the border is bound to bring about dire consequences, including death and enslavement in the worst cases, and ghastly disappointment in the best. Indeed, none of the movies dealing with the question of illegal immigration to the United States ever presents a successful outcome for the illegals. In addition, the political divide does not imply a parity in relations between the two states: while the two police forces are sometimes shown collaborating to stop the human traffic, substantial differences remain in the depiction of the people 'trafficked' (the Mexicans), the people who organise the trafficking (the American coyotes) and the people who have the 'right' to patrol the fence (the American forces).

Going south of the border from the American side is a totally different story: it guarantees adventure, nostalgia, sex and love, and an opportunity to display the American geopolitical 'mission' Seen from the north, the US–Mexico border loses its role as a political boundary and is represented as America's last frontier. It is not a barrier but an open gate. It is the place where it is still possible to enjoy the taste of freedom (lawlessness) that characterised Turner's experience of the frontier. In this case, it is a symbolic landscape, fictitiously constructed by so many narratives that it has become real. Again, it is a gendered landscape, where – in contacts with an Other that is traditionally constructed as inferior and passive – it is still possible to fantasise about a gender and race supremacy that is increasingly being erased in other areas of the national narrative.

CONCLUDING REMARKS

The US–Mexico border is certainly one of the most depicted screencapes in American film. A signifier with more than one meaning, the border has been turned by American popular culture into a powerful icon, framed and reframed in the making of the complex narrative of national Self in its relation with the external Other. From a critical geopolitics perspective, the cinematic border can be defined as a representational practice, the popular form of expression of a geopolitical discourse pertaining to the role and standing of the US relative to its southern neighbour. By analysing this practice, it is possible to reveal the assumptions at the root of this discourse.

These assumptions can be detected in the visual representations of the border and the scripts of movies about the border and the crossing-the-border experience. Via the investigation of more than thirty feature movies, it has been possible to identify several ways of depicting the border: that is – in political geography terms – as a borderland, as a border economic region, as a demarcated boundary and as a frontier. All these different depictions make their own geopolitical assumptions. Together, they form a powerful geopolitical representation of Self and Other.

In this representation, the dual rendering of the US–Mexico border as a closed boundary for the Mexicans and an open frontier for the Americans highlights the United States' double standard in its relations with the rest of the world, in which the United States does not see itself as an equal among equals, but as the champion of the geopolitical mission of civilising and democraticising the less developed Other, on the one hand, and as the more powerful neighbour that can decide who is and is not allowed to enter its territory, on the other.

From this second perspective, the border between the United States and Mexico can also be used as a geopolitical metaphor, symbolising not only the political separation between two sovereign states, but also, on a different scale, the fundamental divide between the Haves and the Have-Nots – that is, between a first world, which is easy to leave but inaccessible from outside, and the rest of the world, which is extraordinarily difficult to leave but easy to enter. Obviously, the same border does not have the same meaning from both sides. At least this is the way things currently stand – although, as science fiction movies such as *The Day after Tomorrow* remind us, to be on the right side of the border, or on the wrong one, is just a question of luck.

NOTES

1. As stated by Daniels, 'national identities are co-ordinated, often largely defined, by "legends and landscapes"; by stories of golden ages, enduring traditions, heroic deeds and dramatic destinies located in ancient and promised home-lands with hallowed sites and scenery. The symbolic activation of time and space, often drawing on religious sentiment, gives shape to the "imagined community" of the nation. Landscapes, whether focusing on single monuments or framing stretches of scenery, provide visible shape; they picture the nation. As exemplars of moral order and aesthetic harmony, particular landscapes achieve the status of national icons'. See S. Daniels, *Fields of Vision. Landscape Imagery and National Identity in England & the United States* (Princeton, NJ: Princeton University Press 1993) p.5. This is true for landscapes that are the product of the works of painters and designers, which are the

focus of Daniels's research, but also for literary and cinematic landscapes. See J.R. Short, *Imagined Country, Society, Culture and Environment* (London and New York: Routledge 1991).

2. M. Shapiro, *Cinematic Political Thought: Narrating Race, Nation and Gender* (New York: New York University Press 1999).

3. C.F. Fox, *The Fence and the River. Culture and Politics at the U.S.–Mexico Border* (Minneapolis, MN: University of Minnesota Press 1999) p.76.

4. R.L. Fregoso, 'Recycling Colonialist Fantasies on the Texas Borderlands', in H. Naficy (ed.), *Home, Exile, Homeland: Film, Media, and the Politics of Place* (London and New York: Routledge 1999) pp.169–92.

5. See Fox (note 3) p.76.

6. Postcards also served this purpose, which, as stated by Fox (note 3) p.77, 'satisfied the public's hunger for images of the war, especially for whose who did not have access to movie theatres'. See in this regard P. Vanderwood and F.N. Samponaro, *Border Fury. A Picture Postcard Record of Mexico's Revolution and U.S. War Preparedness, 1910–1917* (Albuquerque, NM: University of New Mexico Press 1988).

7. See Fox (note 3) p.76.

8. These films depicted the persecution of national and racial stereotypes developed as early as the 1830s, when American popular literature began demonising Mexicans as lazy, stupid and cruel. At the beginning of the twentieth century, silent films also had their input, providing images of Mexicans as bandits attacking white people. See L. Holtzman, *Media Messages. What Film, Television, and Popular Music Teach Us About Race, Class, Gender and Sexual Orientation* (New York and London: M.E. Sharpe 2000) p.225.

9. In this regard, see D. Maciel, *El Norte. The U.S.–Mexican Border in Contemporary Cinema* (San Diego, CA: Institute for Regional Studies of the Californias 1990).

10. A. Horton, 'Reel Landscapes: Cinematic Environments Documented and Created', in I. Robertson and P. Richards (eds), *Studying Cultural Landscapes* (London: Arnold 2003), pp.71–92.

11. A classic contribution on this topic is W. Prescott, *Political Frontiers and Boundaries* (Chicago, IL: Aldine 1987). See also D. Newman, 'Boundaries', in J. Agnew, K. Mitchell and G. Toal (eds), *A Companion to Political Geography* (Malden, MA: Blackwell 2003), pp.123–37.

12. In this essay, I will take into account at least three of the four elements outlined by L. Zonn, 'Tusayan, the Traveler and the IMAX Theatre: An Introduction to Place Images in Media', in L. Zonn (ed.), *Place Images in Media. Portrayal, Experience, and Meaning* (Savage, MD: Rowman and Littlefield 1990) pp.1–5, for analysing the images of places in media: the medium, the place, the collection of individuals who created the image. Discussing audience, the fourth element, is more complex, since common visions associated with a specific place or with a certain landscape either can or cannot be shared by different audiences. From a critical geography perspective, however, we may consider them as the result of a certain geopolitical discourse, locally embedded and historically produced.

13. G. Ó Tuathail, *Critical Geopolitics* (Mineapolis, MN: University of Minnesota Press 1996), and G. Ó Thuatail and S. Dalby (eds), *Rethinking Geopolitics* (London: Routledge 1998).

14. As stated by J. Burgess and J. Gold in their seminal study *Geography, the Media and Popular Culture*(London and Sydney: Croom Helm 1985) p.I, 'Raymond Chandler is as valuable a source as Thomas Hardy, and the *Daily Mirror* has as much to say about the nature of places as the *Geographical Journal*'.

15. In fact, in those years 'a great number of B Westerns were produced in which cowboys vested with the aura of the Old West did battle with problems and villains from the modern world – gangsters, corrupt political bosses … Nazi agents', R. Slotkin, *Gunfighter Nation. The Myth of the Frontier in Twentieth-Century America* (Norman, OK: University of Oklahoma Press 1998), p.273.

16. Ibid. p.276.

17. Since, as J. Limón, *American Encounters: Greater Mexico, the United States, and the Erotics of Culture* (Boston, MA: Beacon Press 1998), p.199, rightly points out, 'the western, with its ritual expectation of the "ride to Mexico" for manly adventure, for sex and drink but also for self-aggrandizement, is fundamentally a genre of imperialism'.

18. L. Engel (ed.), *The Big Empty. Essay on the Land as Narrative* (Albuquerque, NM: University of New Mexico Press 1994).

19. The Cisco Kid was introduced by O. Henry in a series of short stories that were later collected and published in *Heart of the West* (1907). His adventures were portrayed in about 30 movies (the first silent, the last shot in 1994) as well as in a television series that ran for 156 episodes from 1950 to 1956.

20. W. Wright, *Six Guns and Society. A Structural Study of the Western* (Berkeley, CA: University of California Press 1975), and N. Carroll, 'The Professional Western: South of the Border', in E. Buscombe

and R.E. Pearson (eds), *Back in the Saddle Again. New Essays on the Western* (London: British Film Institute 1998) pp.46–62.

21. M. Coyne, *The Crowded Prairie: American National Identity in the Hollywood Western* (London: I.B. Tauris 1997), p.158.

22. There are also comedies in a similar vein. In *Three Amigos* (1986), for instance, some American actors posing as western heroes unwittingly come to the aid of some poor Mexicans.

23. Engel (note 18) p. 227.

24. They die fighting, but not without demonstrating their American superiority, since they know how to handle military technology (in the form of a machine gun) much better than the locals.

25. The other two volumes of McCarthy's trilogy are *The Crossing* and *Cities of the Plain*.

26. *The Mexican* cannot be listed among movies about the border, since Jerry goes to Mexico by plane. It is interesting all the same because it reveals a self-consciousness about stereotyped American attitudes towards their southern neighbour, while being a victim of the same stereotypes it is trying to highlight. For example, two of the characters are shown renting a car when they land in Mexico. Both of them ask for a 'Mexican car', instead of accepting the perfectly acceptable American make offered by the rental firm. Also, when in Mexico the Americans are shown speaking in a very stupid and condescending way – 'I no fuck around, *entiende*?' says one of them to a Mexican. 'Do you have a speech impediment?' is the Mexican's answer. On the other hand, Mexico is portrayed as a crazy country where anything can happen, including dying in a *noche de fiesta* and people speaking with dogs, all in a lethargic society that is apparently sleeping off its own past.

27. See A. Garcia, 'Casting out the serpent: Eroticised Violence and the Mexicana Body in Hollywood Cinema', *Diâlogo*, No.6, Winter/Spring 2002.

28. *The metaphor of the border as a 'magic curtain' is introduced by T. Torrans,*The Magic Curtain: The Mexican–American Border in Fiction, Film, and Song *(Texas Christian University Press: Forth Worth 2002).*

29. S. Vaquera-Vasquez, 'Wandering in the Borderlands: Mapping an Imaginative Geography of the Border', *Latin American Issues*, Vol.1, 1998, section VI.

30. After Mexico's participation in the Second World War, it was no longer the time for openly racist, overtly anti-Mexican movies, such as those produced in the years of the Punitive Expedition. In addition, immediately after the end of the war, the United States launched the so called Bracero Program, under which Mexican workers could be legally hired in America. Notwithstanding this bilateral agreement, many Mexicans did not undergo the bureaucracy required to get legal permission and illegal immigration continued to claim its many victims.

31. R. Kitchin and J. Kneale (eds), *Lost in Space. Geographies of Science Fiction* (London: Continuum 2002).

32. See S. Mains, 'Imagining the Border and Southern Spaces: Cinematic Explorations of Race and Gender', *GeoJournal*, 2004, pp.253–64.

33. Fregoso (note 4) p.170.

34. See Fox (note 3).

35. Fregoso (note 4) p.178.

36. Among the others, see Clark-Jones, Limón (note 17), Fregoso (note 4), Shapiro (note 2), Mains (note 32); and J. Burton-Carvajal, 'Oedipus Tex/Oedipus Mex: Triangulations of Paternity, Race, and Nation in John Sayles's *Lone Star*', in E. Shoat and R. Stam (eds), *Multiculturalism, Postcoloniality and Transnational Media* (New Brunswick, NJ: Rutgers University Press 2003) pp.129–52, W. Somerson, 'White Men on the Edge. Rewriting the Borderlands in *Lone Star*', *Men and Masculinities* 6 (3 January 2004) pp.215–39.

37. See Límon (note 17), 1998, Fregoso (note 4) and Burton-Carvajal (note 36).

38. See Fregoso (note 4) p.183.

39. The notion of 'contact zone' ins introduced by M.L. Pratt, *Imperial Eyes. Travel Writing and Transculturation* (London & New York: Routledge 1992).

40. See Maciel (note 9).

The Demise of 'International Relations': America's Western Palimpsest

MICHAEL J. SHAPIRO
Department of Political Science, University of Hawai'i, Hawai'i, USA

INTRODUCTION: WESTERNS

One of the legacies of the 'western' political theory canon is the myth that the modern state emerged as a historically evolving contract, in which individuals, seeking to avoid getting caught up in a war of 'everyman against every man', assent to a centralised authority. For security reasons, or so the story goes, they participate in a collective sensibility that assembles and legitimates the body politic.[1] Among the more articulate and strenuous challenges to this enduring fable of the social contract is Michel Foucault's. In his lectures at the *College de France* 1975–1976, he challenges those legendary histories, embedded in the political theory canon, that presume that the state is founded by the motivation to avoid war. Contrary to the familiar social contract theories of the origin of states, Foucault argues that state-initiated legalities or jurisdictions cannot be attributed to social consensus; instead, he insists, they have followed in the wake of battles.[2] While Foucault's historical field of reference is the European continent, a similar and compelling case has been made for the United States, whose realisation as a nation-state on the North American continent is attributed by Richard Slotkin to 'the rule of the gun'. Specifically, Slotkin locates America's national consolidation in its Indian wars rather than in the deliberations leading to its foundational charters. And he turns to western films as the genre in which the westward moving frontier of violence has been mythologised to legitimate Euro-America's acts of violent (dis)possession.[3]

However, even classic westerns provide an ambiguous support for the legitimation to which Slotkin refers. For example John Ford's westerns, which created the most familiar Hollywood version of the West, portray a heroic nation-building resolve but also depict a nation-destroying cruelty that threatens the national mythology in which his films participate. Robert Burgoyne's observations about westerns and America's self actualising narratives disclose that threat. Drawing on a remark by the Cherokee artist and writer, Jimmie Durham, who writes: 'America's narrative about itself centers upon, has its operational center in a hidden text concerning its relationship with American Indians…the part involving conquest and genocide, [which] remains sacred and consequently obscured', Burgoyne adds that 'one of the most durable and effective masks for the disguised operational center of the nation-state has been the western, a genre that has furnished the basic repertoire of national mythology'.[4] Although, for the most part, the classic western film genre represents Indians as a temporary impediment to the Euro-American nation-building project, it nevertheless calls attention to a historical epoch in which there is undeniable evidence of an encounter of nationhoods, a period in which, prior to white dominance, there was a process involving Euro-Native American international relations.[5]

Effectively, before Euro America turned Native Americans into a 'race', their encounters with native peoples constituted a period of extended 'foreign policy'.[6] During the period of European colonisation of the North American continent, 'the constructed category 'Indian' occupied the space of the quintessentially 'foreign'.[7] This was also the case from the Native American point of view, as a notable encounter in the American Southwest in 1847 indicates. A Comanche chief, Janamata (Red Buffalo), tried unsuccessfully to instruct a US Army Major, John P. Cremony, about the difference between Euro American reasons of state and his Apache nation's politics of grievance, which had just been activated as an attack on a Mexican town across the Rio Grande, near where Cremony and his troupes were protecting a mining operation on the American side. When Cremony asked Janamata to end his raids, the conversation, in their only shared language, Spanish, was as follows: Cremony tells Janamata that he must recross the river and leave the Mexicans in peace. Janamata says in reply, 'I hear your words and they are not pleasant. These Mexicans are our natural enemies', and, inasmuch as he is familiar with the history of the US–Mexican War he adds, 'they are also your enemies'. Cremony rejoins to the effect that the Mexicans are no longer US enemies and explains that the friend–enemy relationship must be understood in the context of reasons of state, not tradition or historical grievance. To this, Janamata responds that 'your revenge is for yourself. It does not satisfy us for the blood of Comanches slain by Mexicans'. Cremony, backed by superior force of arms, refuses to accept a pluralistic model of war making/peace keeping.[8] This paradigmatic encounter between two different imperatives vis à vis enemies was one among many inter-nation issues that arose before the US government effectively disarmed and displaced the remaining Native American nations and imposed a model of a polity in the form of a fixed and sovereign, territorial state with no space for the more fluid territorial organisation characteristic of many of the Native American nations.

In their construction of Indian territories, the classic westerns tend to bypass the historical trajectory of the Indian's relocation from nations with treaty-making rights and territorial integrity to races, under the aegis of the US government. Instead, they focus on Euro-American national expansion and the dramas accompanying the tensions accompanying white settlement. Nevertheless, although the era of the classic western is now decades in the past, some contemporary films, which return to the same western landscapes, deserve recognition as critical engagements with both the history of Euro-American expansion and the film-nationhood projects involved in the classic westerns. Accordingly, analyses of two such films are the book ends of this investigation, whose middle section treats the way classic westerns, particularly those directed by John Ford, supply insights into the Euro-American geo- and biopolitical 'ethnogenesis', their ethnically privileged and territorially pervasive presence on the North American continent.[9]

I begin with Sean Penn's *The Pledge* (2001) and conclude with Wim Wenders's *Paris, Texas* (1984). Although neither of these films tends to be recruited into the category of the western – the former seeming to be a story about a man losing his mind and the latter about one who regains his – the settings and much of the imagery in both contain subtle references to the western film genre and testify to the violence attending Euro-American 'nation building' (to repeat the euphemism that remains central to the academic discipline of political science in the United States). Temporally, the two films remain in the present; their testimony to events from a violent past is achieved through portrayals of the traces that remain on the landscape. In this respect, their approach to the genocidal assault on Native America (or what euphemism-resistant analysts have called 'nation killing' and 'nation-destroying')[10] is comparable to Claude Lanzmann's monumental documentary on the Holocaust, *Shoah*, which contains no historical footage but rather documents what remains on the landscape and in the memories of survivors and witnesses. Deliberately excluding archival materials, Lanzmann's recording of what remains of a violent, genocidal past in the present is conveyed by images such as 'the overgrown siding at Sibibor [where concentration camp inmates were off-loaded] pecked at by chickens or the scrubby copse that now grows on the site of [the extermination camp] Treblinka's so-called "hospital"'.[11]

THE PLEDGE

I begin with a treatment of *The Pledge*, which, like Lanzmann's *Shoah*, is set well after the genocidal period; it is filmed in a West whose violent past also retains few visible markers on the landscape and in the daily lives of its residents. The far west, Reno-area venue of *The Pledge* is now a well-established region of the contemporary United States. It is a 'region' in the sense that the former 'boundaries on the land', particularly the most dramatic and fluid one – that between Euro-American settlers and 'Indians' – has disappeared. The inter-nation 'frontier', in which Self–Other identity relationships were once subject to contention and negotiation, has been displaced by a more stabilised form of possession and set of identity relations. Whereas in the frontier moment, people were inventing a world, with regionalisation, 'people felt that they were no longer inventing a world but inheriting one'.[12] The discourses associated with the management of the US state's geopolitical prerogatives have left little discursive space for the violent spatial dynamic that preceded national consolidation.

In short, once the inherited world became established, within the dominant geopolitical cartography, alternative modes of social and political life were rendered invisible. As John Agnew points out, the prevailing geopolitical imagination, which was inaugurated in sixteenth-century Europe, remains the primary mode of global visualisation. And, inasmuch as this geopolitical imagination provides but one vision and effaces history by turning time into a singular model of space, the overcoded life worlds of Native American nations, victimised by Euro-American 'nation-building', are difficult to recover.[13] In analysing Sean Penn's *The Pledge* and other films my focus is on imagery in some cases, and alternative loci of enunciation in others, in order to recover what the modern geopolitical imaginary hides. At the same time, I am concerned with showing how a state-centric macro-level politics hides the continuing micropolitical struggles of Native American peoples.

Historically, the relationship between cinema and nationhood has been intimate. In most cases, and in many national venues, film has been supportive of state cultural governance – the practice in which state-oriented cinema constructs homogeneous and coherent national cultures. Elsewhere, I have referred to this phenomenon as 'cinematic nationhood', where film has been involved in the cultural articulation of the nation building and sustaining projects of states.[14] However, as I also note, pointing in particular to 'post-westerns', film is an increasingly critical medium that has tended to challenge the hegemony of the modern state system. In subtle ways, *The Pledge* participates in that challenge.

In *The Pledge*, the legacy of Native America in the area of Reno is not immediately evident. Throughout the film, social relations within the Euro-American settlement (and in the few Anglo-Indian interactions) tales place in the well articulated and settled normative order of Reno and its mountain and lake surroundings (although the filming actually tales place in British Columbia, Canada). Indian presence is sparse, existing for the most part as a behaviourally indistinguishable clientele for small businesses – a gas station in one scene and a diner in another. Briefly, as a plot about an individual character, *The Pledge* treats the events surrounding the long-time Reno police detective Jerry Black's (Jack Nicholson) retirement, which is compromised by a promise he makes to the mother of a murder victim, an eight-year-old girl, whose mutilated body is discovered by a teenage snowmobiler in a snow field outside of Reno.[15] The report interrupts Jerry's retirement dinner party and, ultimately, his retirement. Resisting his department's position that the case is closed, and driven by his promise and his belief that the

Indian (Benicio del Toro), on whom the police pin the murder is not the perpetrator, Jerry works at solving the crime as a freelance investigator. His mapping of a pattern of similar crimes leads him to take up residence in the vicinity where they occurred. Once again, although that vicinity seems merely to be the setting of contemporary, Euro-American daily life; it contains signs of the international encounters whose resolution has integrated the region into a seemingly culturally unified, sovereign United States.

Eventually, as result of local encounters and contingent circumstances, he ends up living with Lori (Robin Wright Penn) and her daughter, Chrissy (Pauline Roberts), who fits the victim profile. After wrongly suspecting a local part-time clergyman, Jerry focuses convincingly on a likely suspect, who has arranged a meeting with Chrissy. But after Jerry and his former colleagues set up a surveillance post, adjacent to the spot of the meeting, the suspect fails to show up (seemingly because he is killed in a traffic accident). The film ends as it begins – with a flashback – with Jerry wandering around in a daze, muttering to himself, a man broken by a combination of desires and contingencies that provoke his investigation, disrupt its consummation, and leave the viewers without a clear answer to whether Jerry is destroyed by self-absorbed delusions or by contingent circumstances.

The film's dramatic tension is maintained by the ambiguities in what drives Jerry's investigation. Although the evidence of a serial killer, and thus the Indian's innocence, becomes increasingly plausible – the murder victim had drawn a picture of a man, who gave her gifts, shortly before the murder, and his body and car in her drawing did not fit the Indian suspect – it also sometimes seems that Jerry's interpretations of his evidence are obsession driven. In particular, the camera shots that focus on Jerry's facial expressions encourage a psychological reading of his quest. Such shots constitute what Gilles Deleuze calls 'affection-images'. They are primarily close-ups of faces at intervals of 'indetermination', moments between what the subject perceives and an action, moments, moreover, when the subject 'feels itself "from the inside"'.[16] These shots, which are the primary vehicles for conveying the psychological register of a cinematic narrative, occur often in the latter scenes of *The Pledge*, as Jerry Black becomes increasingly obsessed with his investigation and increasingly attentive to ambiguous clues.

A key scene in the psychological drama takes place in the office of a counseling psychologist. After Jerry asks her to help him interpret the victim's drawing of 'a porcupine giant' (a tall stranger who gives the child small carved-wooden porcupines), she shifts the interview to an interrogation of Jerry, asking him if he has always been a chain smoker, if he is recently retired, and whether he is still sexually active. As this later part of the session progresses, the camera offers frequent affection-images. Yet while the frequency of affection-images may foreground the psychological level of meaning in the narrative, its mode of presentation has an additional significance. To the extent that the face close-up is 'abstracted from spatio-temporal coordinates', it has the effect of 'undoing space',[17] which in the case of *The Pledge* is a pervasive part of the film's supplemental and largely disjunctive story. But in addition to its exploration of characters, personalities, and individual fates, much of the camera's focus is on the landscape. And, inasmuch as the cuts and juxtapositions encourage seeing landscapes scenes in the context of the scenes containing signs that the West retains a mixed ethnoscape, the many panning shots of the landscape have resonances of earlier international encounters.

Thus, although interiors – primarily offices and small businesses – contain much of the dramatic interaction of the film, time is ever present, for *The Pledge* never abandons its spatio-temporal thematic. Certainly the film is concerned with a character. Jerry Black is situated in time, first as an ageing retiree, then as one partially stymied by the temporal rhythms of police investigations (once a suspect is selected, there is enormous pressure to close the case), and finally as one whose investigative opportunities are affected by seasonal changes (there are several seasonal tableux that are interspersed in the imagistic mechanisms of the story line). But surrounding one man's drama is an institutionalised space, a western venue with a significant temporal trajectory. Reno and its surroundings exist in both ethnic and geopolitical time. The area is not only a consolidated and recently white-dominated region of the West but is also a part of the larger national picture.[18] That Reno is a city in a state in a nation, and is therefore subject to a dense normative hierarchy, is made evident by, among other things, a parade scene in which many national flags and other symbols of a national story (e.g. disabled war veterans) are in evidence. And most significantly, the signs of the process of the whitening (the process of Anglo dominance of what were formerly Indian nation places) are presented abundantly, not only in the few scenes that include Native Americans and in some of the landscape scenes (which include both panoramas and depth of focus shots) but also in the way Anglos are depicted.

With respect to the spatial narrative of the film (or dysnarrative insofar as it is disarticulated with the individual drama), the opening scene is telling. We encounter Jerry Black, initially ice fishing on a lake outside of town. After a distance shot of a lake and mountain snowscape, interrupted only by a small fishing shack, the camera takes us inside the shack and focuses on Jerry's almost empty bottle of scotch. The sequence of shots portrays a historical rather than individual story. A virtually timeless landscape is juxtaposed with a sign of the Euro-American arrival; a European brand of scotch, Glenfiddich, testifies to the origin of part of white America.

And lest the viewer forget the place of alcohol in the story of the western encounter, the film continues to dwell on alcoholic beverages. Several shots of the revelers at Jerry's retirement party, at a downtown ersatz Hawaii bar-restaurant ('The Luau'), foreground whiskey. One is a tracking shot of a tray carrying a cocktail and one is of a police detective dancing with a drink in his hand. The juxtaposition of Euro and Native America that the focus on alcohol implies is deepened by a series of cuts between scenes during Jerry's retirement party. There is a cut from the party to a scene in which an overweight and awkward young boy, whose complexion is ultra white (so white that pink blush marks show, emphasised with several close-up face shots), appears in the deep snow field near the crime scene on his snowmobile. As the camera shifts from the affection-image to a 'perception-image' (where the images are delivered from the boy's perspective),[19] the viewer sees an Indian running through the snow toward his truck. He is carrying the traps with which he catches game (beaver, we later learn).

While the crime story part of the narrative locates the boy as a witness who places the Indian at the murder scene, another more venerable story is being juxtaposed. Shortly after the Indian is arrested and brought in for an interrogation, a framing shot shows a drawing of an Indian caricature made by a police artist, who is viewing the interrogation room on a closed circuit screen. The caricature, an Indian's head with a single feather, which looks nothing like the suspect, articulates with the dismissive perspective of the police, one of whom, the interrogator, derisively calls him 'chief'. And the interrogation scene itself, after which the Indian grabs an officer's gun and kills himself, is arguably an allegory of white America's ambivalent view of Native Americans. In the process of cajoling a confession from the Indian, the interrogating detective rubs him, hugs him, and speaks to him in a soothing, almost lascivious tone. The homoerotic tenor of the interaction is commented on by Jerry, who after watching the interrogation on the closed circuit monitor, says, 'he practically blew him'.

Since the initial, extended contact in the seventeenth century, Euro Americans have located Native Americans in an ambiguous place, at times romanticising them for their alleged qualities of grace, beauty, bravery and natural genius, and at times abhorring them for their alleged primitiveness, savagery, cruelty and guile. For the Reno police, as for the dominant history of Indian–White relations, the abhorrence codes are ultimately privileged, even as the former ones are allegorically played out. Once the pseudo confession is extracted, and the Indian is nearly faint from the pressure, the interrogator, while holding him from falling, says, 'I gotcha, I gotcha' – the first gotcha referring to the support and the second to having achieved a confession. But, of course, there is a third resonance, the historical one in which whiteness has contained Native America, and settlement has dominated the more nomadic habitus of Native America. And within the crime story, the legacy of the containment and disparagement of Native Americans is so well assimilated into the policing habitus (in which both whiteness and closed cases are valued above justice) that his former colleagues strenuously resist Jerry's attempts to seek an alternate perpetrator.

The Indian suspect's (Toby Jay Wadenah) life style receives relatively little coverage. During the interrogation, he mentions that he hunts beaver.

And in the earlier scene, in which he is seen running to his truck by the fat preteen, pink and white snowmobiler, who appears to be too corpulent to run through the snow, his animal traps are visible, dangling from his side. But the White–Indian life style (and habituation to moving in a snow field) juxtaposition is also made evident by the role that fishing plays for Jerry Black. *His* fishing is represented primarily as a leisure sport (especially for retirees), a representation that is underscored when his retirement gift from the police department is a fishing trip to Baja California. The other Indians that appear, once Toby is out of the picture, are simply customers, whose life styles have been assimilated to the white habitus. Some are shown briefly as part of the clientele of the diner where Jerry meets his short-lived family.

However, an especially telling (and doubtless allegorical) Indian–White interaction takes place at Jerry's gas station shop, which he has purchased to be near the pattern of murders he has uncovered. After Jerry fills the tank of an Indian customer's car, he has difficulty getting the credit card machine to record the Indian's American Express card. As he botches the transaction several times, having difficulty making the slide mechanism on the recording machine work, the two men stare awkwardly at each other. A ex-policeman on unfamiliar commercial ground is failing to process smoothly the credit card of a person whose people have been historically discredited (on their former ground). Doubtless, it is not incidental that it is an *American* Express card, which lends additional resonance to the allegorical aspect of the transaction, calling to mind a history of fraught Anglo-Indian transactions and negotiations. Ultimately, despite the intensity of its foregrounded psychological drama and the suspense it generates around its crime story, the haunted land- and ethnoscape that *The Pledge* presents, primarily with its images, which are often disjunctive with the psychological and crime narratives, address a historical crime, the violence attending the Euro-American continental ethnogenesis.

Although North America experienced an undeniable Euro-American ethnogensis, minutely recorded in mainstream histories of the West and in other genres – for example the Broadway musical, *Oklahoma!* (1948) – there are alternative Wests. While, for example, *Oklahoma!* portrays an almost exclusively White ethnoscape, the territory depicted in the musical contained (and still contains) many displaced Native American tribes, a considerable number of 'Negro cowboys' and 26 all-black towns at the turn of the twentieth century.[20] But the most familiar genre participating in the effacement of alternative Wests remains the classic Hollywood western, among which John Ford's are the most notable and enduring.

JOHN FORD'S CINEMATIC WEST

Arguably, no one has been more influential than director John Ford in locating the Anglo-Indian encounters in the West within the dominant American imaginary. But although some of Ford's westerns seem to celebrate the displacement of a Native America 'wilderness' with Euro-American settlements (symbolised, for example, by the evocation of gardens in both *My Darling Clementine* (1949) and *The Man Who Shot Liberty Valance* (1962)), others offer critical perspectives on the heroic versions of Euro America's continental expansion. While some of his best known films unambiguously portray a whitening of the American continent, others resist a celebration of Euro-American expansion; they *think*, not always by virtue of the main story line but through the image sequence, which is at times disjunctive with a heroic, Euro-American nation-building narrative.

Among Ford's less critical versions of an increasingly Anglo-dominated West is his *Stagecoach* (1939), his first film featuring John Wayne (as the Ringo Kid). Although the film does not display the clashes among alternative meaning systems, Euro versus Native American and eastern versus western Euro culture, the former of which is featured starkly in *Cheyenne Autumn* (1949) and the latter in *The Man Who Shot Liberty Valance* (1962), *Stagecoach* is, nevertheless where Ford first makes the landscape itself effectively one of his major characters.[21] As Gilles Deleuze puts it, in this film, as elsewhere, Ford stages 'intense collective moments' that reflect a historical social dynamic in a setting where 'the milieu encompasses the people'.[22] The moments involved in *Stagecoach* are those that Ford creates to select which character types are to become part of the dominant national culture. He tracks the social changes involved in the shaping of a nation in the context of landscape that appears to invite those changes. As Deleuze insightfully puts it: 'For Ford ... society changes, and does not stop changing, but the changes take place in an encompasser which covers them and blesses them with a healthy illusion as continuity of the nation'.[23]

A cinematic version of a story he found in the magazine, *Colliers*, Ernest Haycock's 'Stage to Lourdsburgh', Ford's *Stagecoach* is more or less complicit with the view of the West favoured in *Colliers*. That view is typified by Owen Wisters's *Colliers* stories and is depicted in Fredric Remington's accompanying illustrations (influences which Ford among other directors of classic westerns acknowledged). The West is presented as an evolving social order, that is to become assimilated into the Euro-American, geopolitical and social space.[24] In *Stagecoach* this evolving order is manifested during the long stage coach journey that occupies the primary narrative and image spaces of the film. The occupants of the stagecoach, a disparate and often feuding group of types, are a microcosm of that evolving social order. While constituting a tribute to the historic expansion of a tolerant social democracy, the film's Anglo characters, a southern gentleman gambler, a prostitute, a soldier's pregnant wife, an outlaw, an alcohclic doctor, a liquor salesman and a banker are, despite their lack of social cohesion, a group of types represented, for all their flaws, as part of a Euro American-dominated future. They are represented as destined to displace unreliable Hispanics and dangerous 'savages' who are shown as characterologically unfit to negotiate a shared political order.

While the romantic part of the story, in which the Ringo Kid and the ex-prostitute, Dallas, become a couple, is one dimension of the film's resolution, the more general, historical resolution depicted involves the successful incursion of White society into Indian country. Nevertheless, the film does convey some ambivalence about that success, primarily through its use of irony – for example the exaggerated fright reactions of the travelers when they encounter an Indian woman, who is the wife of a Mexican managing one of the stagecoach stations. For the most part, however, the Indians are simply a menace in the western landscape. Their attack on the stagecoach is repulsed, thanks to the heroics of the Ringo Kid and the last-minute arrival of the cavalry, which is part of the reach of white governance.

The imagery of the landscape and the tendency of Ford's camera to view Indians from a distance accords with Theodore Roosevelt's influential rendering of the western landscape in his monumental *The Winning of the West*, which simultaneously celebrates Euro-American cultural and territorial expansion and evacuates the Native American presence.[25] As the nineteenth-century anthropologist Lewis H. Morgan noted, to make 'Indian life' intelligible, 'their house life and domestic institutions must furnish the key'.[26] In contrast, Ford's Indians, shot for the most part outside, are represented as threats to white domesticity (especially in *The Searchers*). At a minimum, they achieve no ethnographic depth in *Stagecoach* or in subsequent films. However, as Ford's filmic corpus develops, the injustice of the white dispossession of Native American provenances surfaces. In one of his later westerns, *The Man Who Shot Liberty Valance* (1962), Ford's celebration of Euro-American nation building takes a dark turn. Although a soliloquy by the town's newspaper man Dutton Peabody, when he nominates Ransom Stoddard (James Stewart) for the post of territorial representative (the man whose main claim to fame is his alleged shooting of the outlaw, Liberty Valance), emphasises the film's nation-building narrative, the narrative is compromised by the fact that, Tom Doniphon (John Wayne), who lives by the gun, actually shot Valance, not the book-toting lawyer, Ransom Stoddard. Peabody begins with a gloss on the West's past, which dismisses its prior Native American inhabitants as lawless and violent savages. That West contained, he says, 'the vast herd of buffalo and savage red skins roaming our territory with no law to trammel them except the law of survival, the law of the tomahawk and the bow and arrow'. He then refers to the first wave of representatives of white America, who replaced one form of violence, with another, more heroic, albeit now outmoded variety: 'and then with the westward march of our nation came the pioneer and the buffalo hunter … and the boldest of these the cattlemen who seized the wide open range as their personal domain … and their law was the law of the hired gun'. And, finally, he addresses himself to the present need for statehood in order to incorporate the territory into a populist, democratic nation: 'now today have come the railroad and the people, the citizen, the homesteader, the shopkeeper, the builder of cities .... and we need statehood to protect the rights of every man and woman however humble'. Peabody's genealogy of the characters whose presence has dominated the West at different historical junctures sets up the conclusion he seeks, the appropriateness of electing Ransom Stoddard, whom he identifies as one who 'came to us not packing a gun but a bag of law books ... a lawyer and a teacher' and became subsequently 'a man who has become known in the last few weeks as a great champion of law and order'.

But the misrepresentation that is central to Stoddard's identity impeaches the patriotic, nation-building tenor of Peabody's soliloquy. The 'what kind of man' query that Stoddard aims at Liberty Valance early in the film effectively haunts Stoddard's misleading heroic existence and, by implication, America's nationhood as a whole. Alan Nadel puts it well when he refers to Stoddard's 'chronic inability to give authority to his assertions until be becomes the man who shot Liberty Valance ... until he becomes the person he's not'.[27] Ultimately, the reverence for words and books, for which Stoddard is an avatar, is undermined by the moral ambiguities afflicting his identity.

The issue of words and the West arises again in what is Ford's most sympathetic treatment of Native Americans, his *Cheyenne Autumn* (1964), based on a historic, futile trek by what was left of a branch of the Cheyenne nation, as they attempted to defy the US government and leave their arid southwestern reservation to return to their homeland in the Dakotas. In his earlier *Fort Apache* (1948) Ford also treats white injustice toward the Indians, but there, as Tag Gallagher points out, the dramatis personae are White, never Red; Ford's focus is on 'the traditions and community values that render otherwise decent individuals into willing agents of imperialism and genocide'.[28] But in treating the Indian with more ethnographic depth in *Cheyenne Autumn*, Ford emphasises the Cheyenne's inability to have their words count and includes scenes in which they have decisive, within-nation conversations about their options. The film foregrounds the disjuncture between White and Red systems of intelligibility and, ultimately represents the Euro-American victory as not simply an example of superior fire power but also as a discursive one. As it is put in the film, 'it is white words, white language that have been our potent weapon against Indians'.

Ford's ambivalence toward the victory of White words in the West in *Cheyenne Autumn* articulates with his more general ambivalence toward the White American continental ethnogensis, which is most evident in what is arguably one of the most significant twentieth-century films, Ford's *The Searchers* (1956), in which, for the first time, Ford's West becomes more complicated. Here, the main character, Ethan Edwards, played by John Wayne, who is unambiguously heroic in *Stagecoach* as The Ringo Kid (saving the stage's passengers from an attack by menacing Indians), becomes both hero and anti-hero. On the one hand, he is an exemplar of heroic masculinity, wise in the ways of his adversaries and protective of those with whom he is associated. But, on the other, he is an Indian-hating racist.

Certainly *The Searchers* partakes of the traditional Euro-American patriotism, inaugurated in the silent westerns and continued in the classic ones. Virginia Wright Wexman describes well the patriotic sentiments conveyed in the historical trajectory of westerns that feature Euro-American settlers encroaching on Indian, who are menaced by Indian attacks and saved by the US cavalry: 'patriotic motifs overwhelm the sense of violence and exclusion implicit in such scenarios by appealing to a mythology of American national origin centred on the sanctity of the image of the family farm'.[29] *The Searchers* opens with a shot from inside the family farm out toward a vast, open landscape, from which Ethan Edwards is approaching 'the family farm'. However, the details of the family drama in the film – it is implied that Ethan has had a romantic attachment to his brother's wife – are less important than the historical context and social setting. Ford's film maps the expanding world of post-Civil War America and reenacts the encounters among White settlers, Spanish aristocrats, and Indian nations, shortly after Texas had become independent.

The film's historical moment is right after Texas had violently extracted itself from Mexican control and was in a period of ongoing war with Comanches (who were not defeated until the Red River War of 1874–1875). In the film, a Comanche war party led by Scarface kills Ethan's brother and wife, rapes, mutilates and kills their eldest daughter Lucy, and carries off Debbie, the younger daughter. The episode precipitates a captivity narrative (one of the oldest modes of American literature). The main searchers are Ethan and young Martin Pawley, a part Cherokee, whose Indianness is constantly disparaged by Ethan during their quest to find Debbie and exact revenge for the massacre. The search proceeds for five years, allowing Ford to present the articulation of generational and historical time. Ethan's paternal relationship with Martin (which he resists until making his will near the end of the film) sets a background for Ethan Edwards's anti-miscegenation fixation. Although Ethan ultimately resists the impulse, he has sworn to kill Debbie when he finds her because she has become an intimate of a Comanche and, in his words, is 'no longer white'. Although Ford's film is somewhat more ambivalent about the ethnogenesis of the Euro-American state than Cooper's novels, he offers a similar resolution, put into the mouth of Mrs Joergenson, the mother of Martin Pawley's 'intended', Lucy. She remarks at one point that the West will have a White future, but perhaps not until 'all their bones are in the ground'.

Nevertheless, Ford's film is not simply a narrative of White civilisation displacing savage Indianness in the West. As is the case with Ford's films in general, a strictly narrative account misses the play of opposing forces and world views that clash in the film. Indeed Ford's film anticipates the 'dysnarrative' aspect of later, so-called 'experimental cinema';[30] *The Searchers* contains a resolution-inhibiting encounter between two different orders of coherence, the narrative and the structural. While in the narrative register, a resolution occurs because Ethan decides to bring Debbie home rather than killing her, structurally much of the imagery – stark juxtapositions between dark domestic interiors and a wide, seemingly untamable landscape – suggests that the space is too large to afford easy incorporation into the cultural habitus of any one group. Moreover, there is no clear dividing line between Euro and Native Americans. Certainly Euro-American practices of exchange are juxtaposed to Native American trading practices, most notably the difference between the Double Eagle coins that Ethan throws around and the trade that Martin effects for a blanket (while he is unaware that he is also acquiring a wife). But Ethan and his adversary, Scarface, operate with similar revenge motifs. And Ethan, despite his articulated contempt for Indians, displays many instances of Indianness, for example speaking Comanche, desecrating a corpse to compromise its spiritual future, and scalping the dead Scarface. And the different groups are also similar in their violence. The scene of Ethan's family's burned farmhouse and dead bodies is parallel to a scene of a burned Indian village with dead bodies, courtesy of the Seventh Cavalry.

Most significantly, the antagonisms and boundary protection exhibited by Ethan Edwards is juxtaposed to that of a searcher who is seemingly peripheral to the main search – old Mose Harper, the man who, without even trying, actually finds the missing Debbie (and whose various appearances constitute a dysnarrative). Although the humorous antics of Mose tend to be treated by viewers as mere comic relief (for example, while he and the other Texas Rangers are under fire during a ferocious Indian attack, he exclaims, 'thank thee Lord for what we are about to receive'), Mose can be seen instead as one who operates outside of the racial-spatial order and its attendant antagonisms. While controlling territory, seeking revenge, and, in general, having one's ethnos dominate the landscape is what drives Ethan and Scarface, Mose is unimpressed with boundaries and enemies. In one scene, when chasing Indians is on the agenda, he begins an Indian dance rather than grabbing a weapon. And when he appears at each key juncture in the story, first at the farmhouse when the Rangers form a posse in search of stolen cattle, then as he returns to see the farmhouse burned, and toward the end when he is asked to report on Debbie's whereabouts, a rocking chair is central to the scene. In the first, he sits in a rocker inside the house and, on taking his leave, thanks his hostess for the chair (while the others mention only the coffee).

In the second, he finds the rocker outside the burned-out farmhouse, and while the others are dealing with their anger and grief, he sits down to rock. Finally, in the third, he offers the information to Ethan about how to find Debbie in exchange for a promise of his very own rocking chair. Rather than merely exhibiting a demented and lazy loafer, through the character Mose, Ford's film offers a different kind of life world, one in which the main desire is to abide without ethnic enmity and territorial displacement. For this reason among others, the film cannot be easily enlisted along with those texts that warrant or celebrate the nineteenth-century version of Euro-American imperial expansion. The exceptional ambiguity evinced in *The Searchers* and the criticism of Euro-American imperialism implied in *Cheyenne Autumn* are important for subsequent critical treatments of the Euro-American ethnogenesis, particularly the latter film's emphasis on the dominance of White words in winning the West. The scenes of a vast expanse in *The Searchers* and the insight about a cognitive form of imperialism in *Cheyenne Autumn* have a phantom presence in Wim Wenders's *Paris, Texas*, which contains numerous filmic references to both. Despite how oblique those references are, *Paris, Texas* articulates Burgoyne's above-noted insight about 'the disguised operational center of the nation-state ... [its] hidden text concerning its relationship with American Indians'.

PARIS, TEXAS

There is a remarkable parallel between the opening in Ford's *The Searchers* and Wenders's *Paris, Texas* Recalling the opening scene in *The Searchers* when, after a song about what makes a man wander that accompanies the opening credits, the first images are of a man riding in from a vast panorama in Monument Valley, Wenders's film begins with a man, Travis (Harry Dean Stanton), wandering over a large western landscape, close to the Mexican border. But unlike Ethan Edwards, who has a lot to say, Travis is silent; he says nothing after being treated by a doctor, who treats him after he collapses while trying to find something to drink at a remote gas station/convenience store. And seemingly having wandered 'beyond the frontiers of language',[31] he remains speechless for some time, after his brother Walt (Dean Stockwell) is summoned from California to fetch him.

Although some of the film's narrative can be contained within a Freudian family story, (particularly one oriented around patriarchal violence), the implied historical narrative of the plot, along with the dysnarrative, disjunctive effects of the landscape images, tells a political rather than psychological story.[32] Travis and his brother begin their journey to California without dialogue and with significantly different relationships to language. That Walt is the one who summons Travis back into the world of language is paradoxical. A prototypical middle-class man, Walt's speech consists of unremarkable platitudes, and his company makes roadside billboards. Apart from the banality of the language, both oral and visual, with which Walt is associated, the West to which he belongs evokes Ford's above-noted insight into the words that have won the West, but with a decidedly Wenders inflection.

Wenders's West is '*the* place [in America] where things fall apart'.[33] His West has been abandoned; it is a place where 'civilisation simply passed through', where instead of a lot of people 'in the middle of a desert you come across a road sign reading: 375 Street'.[34] Ultimately, as Wenders puts it, the West holds 'names and writing ... a lot of signboards, cinema facades, billboards half worn away by the elements, already falling apart'.[35] Within such a version of the West, *Paris, Texas's* Walt is an exemplar of Euro America's contemporary habitus. Instead of the settlers's homes, around which the action in Ford's films take place, there are signs and billboards: 'a man will paint a big sign first of all, and then the sign counts for more than the building'.[36] And Los Angeles, the destination of Travis and Walt's journey, for it is Walt's place of residence, represents the ultimate in an ersatz relationship with place (where representations trump built structures). When Wenders's own western journey ends in Los Angeles, he notes 'I was walking around in downtown Los Angeles, and all of a sudden down this dead end I saw a painted wall with its make believe windows and a dainty awning'.[37]

Much of the imagery caught by Wenders's still camera in his western journey is central to his cinematic West in *Paris, Texas,* in which many of the shots exhibit America's contemporary name-dominated habitus. The dominating imagistic aspect of the film's focus on signs is reflected in the location-scouting stills, taken before the filming began, which include motel signs, road signs, abandoned café buildings with names in large lettering and, most tellingly, a small almost effaced sign in the midst of an empty desert tract, which reads, 'Western World Development Tract 6271'.[38] Rather than seeing America's 'world of names' as merely a commentary on a contemporary rift between person and place, *Paris, Texas* offers hints of a history of violent displacement that the naming-dominated White American settlements overcode. The words that won the West no longer seem to need those who vocalise them!

The naming-preoccupation in Wenders and Shepard's screenplay is likely influenced by the Austrian novelist, Peter Handke, for the expressions 'world of names' and 'places of names' appear often in Handke's novelette, *The Long Way Around*, which features a contrast between the Native-American mode of dwelling in Alaska and the Euro-American mode of dwelling is cities such as Berkeley (where apartment buildings have 'colonised woods'), Denver and New York. In this latter mode, for example, the Northern Lights, which for Native Americans has cosmological significance, is reduced to the name of a street, 'Northern Lights Boulevard' in a California city.[39] Foregrounding the 'violence of representation'[40] inherent in such naming practices, the film narrative is concerned with two different levels of violence.

At an individual level, Travis, once he is summoned back into language, recovers the memory of his violence toward his wife, Jane. At the end of his personal search, which is consummated when he restores the relationship between Jane and their son, Hunter, he is able to verbalise his past violent acts and take responsibility for them. At another level, however, the film ponders Euro America's collective violence, primarily through images rather than with its major narrative thread. When Travis returns to his son, Hunter, after a long encounter with his wife, the film cuts to an alley where the worlds 'RACE, BLOOD, LAND' are written on a wall. At roughly the same time, the camera shows a large head of a Native American woman, as part of a mural on the wall, and an African-American version of the Statue of Liberty, on a wall nearby. And yet another aspect of the history of imperial violence is thematised. When Travis finally converses with his brother, Walt, we learn that they are Hispanic on their mother's side.

> Travis: Do you remember how Momma's very first names was?
> Walt: Mary.
> Travis: No, I mean her…before she found Daddy.
> Walt: Oh, her maiden name.
> Travis: Yes.
> Walt: You don't remember her maiden name?
> Travis: No.
> Walt: Sequin
> Travis: Sequin. Spanish?

At an individual level, the exchange is about Travis's faulty memory, which accords with his other, temporary memory deficits. But at a collect level, it can be connected to America's forgetfulness of its considerable Spanish heritage. It becomes evident that the contemporary American geopolitical imaginary is as forgetful of the demise of Spanish America as it is of Native America. The geopolitical map of the United States fails to record the fate of Mexican Americans, who, during the period in which Anglo conquerors took over California, Texas and the rest of Spanish America, 'found themselves inside the United States [as] foreigners on their own land'.[41] America's western palimpsest is written over its erased Hispanic as well as its Native American markings.

Subsequent to the conversation that restores his Mexican heritage, Travis sings a melancholy Spanish song while in Walt's home, and later he renews his familiarity with this part of his heritage in a conversation with his brother and sister-in-law's Mexican maid, Carmelita, who asks him identity questions to help him select a mode of dress and a body language. One ready-to-hand interpretation of the mixture of cultural and national codes – including the disjunctive Paris–Texas, a German doctor who treats Travis after he collapses, Walt's wife, who is French, and the Spanish maid, Carmelita – is to evoke the idea of cultural conflict,[42] which, over the course of his career has been one of Wenders's cinematic signatures (for example it is central to his *An American Friend*). From a critically oriented, political perspective, however, it is more apposite to invoke the concept of *critical translation*. Recalling Walter Benjamin's insight that the translator is coming to terms with her/his own national–cultural–linguistic self-understanding while making sense of a foreign text, one can read Wenders's approach to Euro America's violent past as not only a recovery of American history but also as a reflection on Germany's Nazi past as well as on the arbitrariness of geopolitical boundaries.

Wenders's film can therefore be seen as evoking the comparison (treated above) between the genocidal assault on Native America and the Shoah. Again, Handke's *The Long Way Around* provides a proto-text. His main character, the Austrian geologist, Valentin Sorger, discovers the violence of names on the American continent while at the same time seeking to come to terms with the violence of the Nazi period in Germany and Austria. Attempting, through his simultaneous reading of America and Europe, to evince a 'science of peace' in the present, Handke's Sorger discovers that he must first free himself from being an unreflective part of his national patrimony, from being 'the faithful replica of death-cult masters'.[43] Like Handke, Wenders effects a simultaneous treatment of Nazi German and Euro America's racial violence – dramatically through the above noted slogans and figures on the wall, but he also through one of the film's characters. *Paris, Texas* evokes the violence of Germany's past through the German- accented doctor, who treats Travis before the point where he recovers from his collapse and 'falls literally into the world of language'.[44] The doctor turns out to be corrupt and greedy; his demand of a large payment from Walt as a ransom for Travis is another aspect of Wenders's critical look backward at the biopolitics of Germany during the Nazi period.

Ultimately, Wenders' *Paris, Texas* contains a strong narrative similarity to Ford's *The Searchers* inasmuch as both searchers, Ethan and Travis, remain loners at the ends of the stories. But in *The Searchers* Ford, as in his other westerns, displays an ambivalent position on myth, seeing it as both destructive and necessary ('print the legend' as he says through a newspaper editor in *The Man Who Shot Liberty Valance*). His films have continued to produce a West that by 1893, as Slotkin notes, had been historically 'closed', as 'a geographical place and a set of facts requiring historical explanation'. Instead, it is ready-to-hand for people to use as an unproblematic mythic West, 'a set of symbols that *constituted* an explanation of history'.[45] By contrast, Wenders is unambivalently hostile to national myth, both Germany's and America's. And, unlike Ford, he is unambivalently hostile to violence, which for Wenders has no heroes. As a result, in *Paris, Texas*, he restores the West as a place whose history must be reopened, as a set of symbols requiring critical explanation.

The implications of Wender's (and Sean Penn's) cinematic reopening the West to critical reinterpretation can be reinforced with a consideration of a politically pregnant episode of resistance by remnants of the Oglala Sioux nation at what came to be known as 'Wounded Knee II', a resistance that, among other things, rendered unstable the geopolitical map of South

Dakota, which, I should note, still contains an extensive Native-American, albeit tribalised, presence. In turning to recent social movements and a reading of Mount Rushmore, as it was initially conceived and as it was subsequently reinflected in film by a Native American director, I am introducing a challenging spatial practices and a critical loci of enunciation that reveal a history of wrong within the dominant geopolitical imaginary. A treatment of the recent social movements, which resists the discourses on law enforcement through which they were portrayed in official and mainstream media and substitutes a critical historicising perspective for the juridical one that has reinforced Euro America's geopolitical prerogatives.

CONCLUSION: CONTESTED SITES IN 'THE MOUNT RUSHMORE STATE'

The map of South Dakota retains stark aspects of both Euro-American imperialism and Native-American presence. Arguably, no national monument is more thoughtlessly intrusive and forgetful about the history of imperial violence in the West than Mount Rushmore, which stands as an icon to what Joseph Conrad called 'geography triumphant' in his gloss on the geography–imperialism nexus.[46] Ceclia Tichi's commentary, evinced upon viewing the monument is especially apropos:

> In August 1965, Mt. Rushmore viewed through binoculars was not a contested site, at least not to a mainstream white. Those lenses did not show a sculptural ideological message of white male authority over minority groups or women. Nor did the huge heads show the extent to which Mt. Rushmore could be considered an imperialist project, since white mainstream America knew nothing of the monument's violation of the sacred ground of the Oglala Sioux (and might well have been indifferent had it known, since the Native American rights movement had not yet organized its civil rights initiatives).[47]

However, subsequent to Tichi's remarks, the 'civil rights initiatives' *had* developed and had even escalated well beyond the issue of rights. Native American activists began to recode the landscape, seeking to recover their coherence as a territorially based nation in South Dakota. Specifically, on 27 February 1973 roughly 200 members of AIM, the American Indian Movement, began an occupation of the village of Wounded Knee, South Dakota that was to last almost three months. Rather than merely a rights initiative, the occupation was an attempt to reestablish Native American *sovereignty* and thereby challenge the racial identity model within which their territorial and historical nationhood is denied. The occupiers declared themselves an independent state at the historical site of the 1890 massacre of approximately

300 Indian men, women and children by US soldiers, under the command of General Crook.

The political force of the occupation was thus both temporal and spatial. Temporally, the choice of the historical landscape of massacre, which has since had powerful collective resonances for Native Americans, constitutes an exemplary case of a political 'subjectification',[48] an action or movement through which a group that has had no legitimate place within the political order have, in Jacques Ranciere's terms, 'made themselves of some account' by 'placing in common a wrong'.[49] But while the political impact of selecting Wounded Knee operates within a temporal or historical register, the actual tactics employed during the extended occupation functioned spatially; on 10 March, 'The Wounded Knee occupiers declared a new state: The Independent Oglala Nation (ION)', and this refiguring of the Native American geopolitical status changed the significance of Wounded Knee's boundaries: 'the boundaries that defined this contested site were now declared "borders", whose status was to be aggressively maintained by a "border patrol"'.[50]

The sudden reconfiguration of the meaning of the Indian-white boundary – in D'Arcus's terms, 'a quite concrete reconfiguration of space'[51] – is a form of political resistance at a material, territorial level. The whitening of the South Dakota landscape, dramatically registered by the Rushmore faces, has also been challenged cinematically. Among the cinematic challenges is an early, oblique one articulated in Alfred Hitchcock's *North by Northwest* (1959), which ends with a dramatic fight between the protagonist, Roger Thornhill (Cary Grant), and his attacker on the faces of the Rushmore sculptures. To appreciate the implicit critique one must recognise the way Hitchcock treats landscapes. As one commentator has noted,

> Hitchcock's camera typically only begins by enacting a survey of a seemingly natural scene. Eventually, as the filming proceeds, it becomes evident that there is a perverse element in the landscape … the film's movement invariably proceeds from landscape to stain, from overall shot to close-up, and this movement invariably prepares the spectator for the event.[52]

Thus, for example, in *North by Northwest*, a biplane crop duster is fogging the ground in an area where – a bystander tells Roger Thornhill – there are no crops to dust. And in the final scene, the Rushmore faces clearly disrupt an otherwise natural setting.

If we keep the idea of the stain on the landscape in mind, we can heed the symbolic challenge evinced in Scott Eyre's film *Skins* (2002), a more direct treatment of Indian displacement. The film's primary narrative is a familial story about Oglala Sioux who live on a squalid reservation in the shadow of Mount Rushmore. But the story contains a more collective, historical level; it ends with a gesture that inter-articulates Native America's reaction to their diminished presence with a Hitchcockian cinematic conception of the perversities and stains on the landscape. An Oglala policeman, Rudy, in mourning for his deceased brother, Mogy, fulfils his promise to his brother, who wants him to attack the Rushmore statue of George Washington. He mounts the attack by heaving an open five-gallon can of red paint over Washington's head. One of the film's last images shows the promise fulfilled while, at the same time, expressing enduring Oglala sentiments about their displacement and the whitening of their landscape. It shows a red stain running from the bridge of Washington's nose to his chin.

To return to my inaugurating theme: while it is unlikely that the 'international relations' that characterised the American continent will be restored, contemporary Native America is asserting itself. Through its writing, music and cinema, among other genres of expression, it is becoming increasingly visible and voluble. Competing with the historical regionalisation of the West, which, as I noted, substituted fixed geopolitical boundaries for a more fluid, negotiated multinational spatial encounter, is a respatialisation and a reintroduction to the consequences of what was an institutionalised silence.

NOTES

1. It is in Thomas Hobbes's *Leviathan* that the quoted phrase exists. See Thomas Hobbes, *Leviathan*, ed. Richard Tuck (New York: Cambridge University Press 1991) p.88.

2. See Michel Foucault *'Society Must Be Defended': Lectures at the College De France 1975–1976*, trans. David Macey (New York: Picador 2003).

3. Richard Slotkin, *Gunfighter Nation: The Myth of the Frontier in Twentieth Century America* (New York: Atheneum 1992).

4. Robert Burgoyne, *Film Nation: Hollywood Looks at U.S. History* (Minneapolis, MN: University of Minnesota Press 1997), p.48.

5. See for example Cheryl Walker's *Indian Nation* (Durham, NC: Duke University Press 1997), where she notes (p.4) that not only was 'America' an encounter of alternative nationhoods (for Native Americans had not understood nations as the Euros had come to define them'), but also that America emerged as a political entity as a result of significant Native American participation in discussions of national identity.

6. For a treatment of this shift, see John Borneman, 'American Anthropology as Foreign Policy', *American Anthropologist* 97/4 (December 1995) pp.659–668.

7. Ibid. p.665.

8. John C. Cremony, *Life Among the Apaches* (New York: Indian Head Books 1991) pp.13–14.

9. I am using the term ethnogenesis in the sense in which it is used in William Boelhower, 'Stories of Foundation, Scenes of Origin', *American Literary History* 5/3 (Fall 1993) p.392.

10. See Pierre L. Van Den Berghe, 'The Modern Nation State: Nation-Builder or Nation-Killer', *International Journal of Group Tensions* 22/3 (Fall 1992), pp.191–208; Walker Connor, 'Nation-Building or Nation-Destroying? *World Politics* 24/1 (January 1972), pp.319–55.

11. The quotations are from Barry Langford , 'You Cannot Look at This: Thresholds of Unrepresentability in Holocaust Film', *Journal of Holocaust Education* 8/3 (Winter 1999), p.29.

12. The concepts (of frontier and region) and quotations are from William Cronon, George Miles and Jay Gitlin, 'Becoming West', in Cronon et al. (eds), *Under and Open Sky: Rethinking America's Western Past* (New York: W.W. Norton 1992) pp.18–23.

13. See John Agnew, *Geopolitics: Re-Visioning World Politics,* 2nd ed. (New York: Routledge 2003).

14. Michael J. Shapiro, *Methods and Nations: Cultural Governance and the Indigenous Subject* (New York: Routledge 2004) p.142.

15. This part of the plot follows that of the novel by Friedrich Durrenmatt (which treats the murder of a young girl in one of Switzerland's eastern cantons), the prototype for the film script; Friedrich Durrenmatt, *The Pledge* (New York: Alfred A. Knopf 1959).

16. Gilles Deleuze, *Cinema 1*, trans. Hugh Tomlinson and Barbara Habberjam (London: Athlone 1992) p.65.

17. Ibid. p.121.

18. The 'American *Indian* population of Nevada, which was almost half of the 16,000-plus total in the early 1860s (see<http://dmla.lib.nv.s/docs/nsla/archives/political/historical/hist18.htm>) is now, in the Reno area, which is representative, slightly over 1 per cent (see <http://dmla.lib.nv.us/docs/nsla/archives/political/historical/hist18.htm>).

19. The 'perception-image'is another of Gilles Deleuze's categories in *Cinema 1* (note 16) pp.66–8.

20. The so-called 'civilized tribes' from the East, who were removed to Oklahoma, brought black slaves and black relatives with them ('thousands of Negroes were neighbors or slaves of the Five Civilized Tribes'), and many became cowboys. See Philip Durham and Everett L. Jones, *The Negro Cowboys* (New York: Dodd, Mead & Co. 1965) pp.18–19. As for the all-black towns, novelist Toni Morrison discovered their history in Oklahoma towns while researching her novel *Paradise*. As she notes, 'I was looking at the book of photographs *Ghost Towns of Oklahoma*', and noticed that 'it scarcely mentions any of the black ones' (quoted in 'Morrison's West', *Vanity Fair* 450 (February 1998) p144.

21. In his biography of Ford, Scott Eyman states that from his earliest films, Ford was already 'making the landscape a character'; see his *Print the Legend*, (New York: Simon Schuster, 1999) p.73.

22. Deleuze (note 16) p.146.

23. Ibid., p.148.

24. For an analysis of the Cooper–Wister contrast and its influence on westerns, see John Cawelti, *The Six-Gun Mystique Sequel* (Bowling Green, OH: Bowling Green University Popular Press 1999) p.66.

25. See Theodore Roosevelt, *The Winning of the West* (NY: G.P. Putnam's Sons 1889).

26. Lewis H. Morgan, *Houses and House-Life of the American Aborigines*, ed. Pau Bohannan (Chicago, IL: University of Chicago Press 1965 [1881]) p.xxiv.

27. Alan Nadel, *Containment Culture: American Narratives, Postmodernism, and the Atomic Age* (Durham, NC: Duke University Press 1995) p.195.

28. Tag Gallagher, 'Angels Gambol Where They Will: John Ford's Indians', in Jim Kitses and Greg Rickman (eds) *The Western Reader*, (New York: Limelight, 1998) p.273.

29. Virginia Wright Wexman, 'The Family on the Land: Race and Nationhood in Silent Westerns', in Daniel Bernardi (ed.), *The Birth of Whiteness* (New Brunswick, NJ: Rutgers University Press 1996) p.129.

30. On the anti-diegetic aspects of experimental cinema, see Andre Gardies, *Le Cinema de Robbe-Grillet: Essai semiocritique* (Paris: Albatross 1983).

31. The phrase belongs to Peter Handke in *A Writer's Afternoon* trans. Ralph Manheim (New York: Methuen, 1991) p.3.

32. Certainly Sam Shepard, who wrote the screenplay, is known for his psychological, Freudian-oriented plots. But at the same time his stories and vignettes explore historical landscapes. See, for example, his western musings in *The Motel Chronicles* (San Francisco, CA: City Lights 1982), which contains imagery that anticipates aspects of his screenplay for *Paris, Texas*. For a treatment that offers a psychoanalytic reading of the film, see Donald L. Carveth, 'The Borderland Dilemma in *Paris Texas*: Psychoanalytic Approaches to Sam Shepard', *Psyart: A Hyperlink Journal for Psychological Study of Arts*, at: <http://www.clas.ufl.edu/ipsa/journal/1997_carveth01.shtml>.

33. Conversation with Alain Bergala in Wim Wenders, *Written in the West* (Munich: Shirmer 1987) p.11.

34. Ibid.

35. Ibid., p.9.

36. Ibid. p.13.

37. Ibid. p.16.

38. The 'location scouting' stills can be seen in Chris Sievernich (ed.), *Wim Wenders – Sam Shepard: Paris Texas* (Nordlinger: Greno 1984) pp.98–9.

39. Peter Handke, *The Long Way Around*, one of three novelettes in *Slow Homecoming* (New York: Farrar/Straus/Giroux 1985) p.76. Both authors of the screenplay of *Paris Texas*, Wenders and Shepard, are familiar with Handke's novels. Wenders worked with Handke on a screen play for one of

Handke's earlier novels (*The Anxiety of the Goaltender During the Penalty Kick*), and Shepard reviewed *Slow Homecoming* in the *New York Times Book Review*.

40. This expression and its implications are well developed in Jacques Derrida's 'Violence and Metaphysics', in *Writing and Difference*, trans. Alan Bass (Chicago, IL: University of Chicago Press 1978).

41. Joy M. Lynch, '"A Distinct Place in America Where all *Mestizos* Reside": Landscape and Identity in Ana Castillo's *Sapogonia* and Diana Chang's *The Frontiers of Love*', *Melus* 26/3 (Fall 2001), p.120.

42. This is the interpretive resort of Phillip Kolker and Peter Beicken in their excellent reading of the film, 'Paris, Texas: Between the Winds', in P. Kolker and P. Beicken, *The Films of Wim Wenders: Cinema as Vision and Desire* (New York: Cambridge University Press 1993) p.125.

43. Handke, *Slow Homecoming* (note 39) p.114.

44. This apt expression is in Kolker and Beicken (note 42) p.120.

45. Quotations from Slotkin (note 3) p.212.

46. Joseph Conrad, 'Geography and Some Explorers', in Joseph Conrad, *Last Essays* (New York: Doubleday, Page & Co. 1926) pp.1–21.

47. Cecilia Tichi, *Embodiment of a Nation: Human Form in American Places* (Cambridge, MA: Harvard University Press 2001) p.2.

48. See Jacques Ranciere, *Disagreement*, trans. Julie Rose (Minneapolis, MN: University of Minnesota Press 1998).

49. Ibid. p.27.

50. The quotations are from B. D'Arcus's excellent analysis of the spatial strategies of the Wounded Knee insurgency: 'Contested boundaries: Native Sovereignty and State Power at Wounded Knee, 1973', *Political Geography* 22/4 (May 2003) p.416.

51. Ibid., p.421.

52. Pascal Bonitzer, 'Hitchcockian Suspense', in Slavoj Zizek (ed.), *Everything You Always Wanted to know About Lacan … But Were Afraid to Ask Hitchcock* (New York: Verso 1992) p.23.

Frames of Reference on the Geopolitical Stage: *Saving Private Ryan* and the Second World War/Second Gulf War Intertext

ANDREW CRAMPTON
Department of International Studies, Nottingham Trent University, Nottingham, UK

MARCUS POWER
Marcus Power, School of Geographical Sciences, University of Bristol, Bristol, UK

INTRODUCTION: RECOVERING A SENSE OF THE 'GOOD WAR'

The last decade has seen a remarkable number of Second World War (WWII) films produced in Hollywood that at times seem part of a more general WWII nostalgia industry. Among the high-budget Hollywood productions taking WWII as their subject are *Saving Private Ryan* (1998), *The Thin Red Line* (1998), *U-571* (2000), *Windtalkers* (2001), *Pearl Harbour* (2001) and *Hart's War* (2002). The prevalence of WWII as subject matter suggests a wider process of recovery at work, a process of recovering something lost. In the aftermath of the end of the Cold War and the 1991 Persian Gulf War Pollard suggests that the increasing focus 'on the great achievements of World War II' in Hollywood is part of a process of 'recover[ing] a sense of the "good war" that might legitimate expanded American military adventures within the New World Order'.[1] Of course, WWII is not the only war restaged by Hollywood in recent years – these films only form part of a re-invigorated film combat genre. Other wars from Vietnam (*We Were Soldiers*, 2002), Yugoslavia (*Behind Enemy Lines*,2001), and Somalia (*Black Hawk Down*, 2001), among others, have all positively represented US international military actions. Nevertheless, the subject of WWII has proven exceptionally popular among combat films in recent years.

The recovery of WWII as a frame of reference for a good or just war has also been prevalent in practical geopolitical discourse particularly from the 1991 Persian Gulf War[2] through Bosnia[3] and Kosovo to the 2003 Gulf War. This period has also seen the emergence of what Der Derian terms 'virtuous war' which refers to a coming together of the virtual and virtuous in attempts to 'effect ethical change through technological and martial means'. 'At the heart of virtuous war' is the capability to 'actualize violence from a distance – *with no or minimal casualties* ... promot[ing] a vision of bloodless, humanitarian, hygienic wars'.[4] For Der Derian, the international intervention in Kosovo in 1999 marked the first properly virtuous war where 'the virtue of the conflict could be measured by the low casualties, the discriminate use of violence, and the application of international norms'.[5] His arguments, however, principally examine the virtuality of virtuous wars that has seen warfare increasingly conducted away from the theatre of operations in command centres and through hi tech wargames and conflict simulations. While Der Derian locates the virtue of contemporary conflict in the supposedly low number of casualties we would like to extend his analysis and argue that contemporary virtue is partially produced through the discursive twinning of contemporary conflict with the Second World War and the representation of warfare as fought for just and humanitarian reasons.

With the spectre and mythological power of WWII looming large in narrations of contemporary conflict it is important to attend to the narratives and myths of WWII written in cinema over the last decade, and also how these myths become twinned with the staging and representation of contemporary conflict. It is beyond the scope of this essay to examine in detail all the WWII films produced in the last decade but we want to focus here on the most successful of these films in box-office terms. *Saving Private Ryan* (*SPR*) was the most popular war movie of the last decade and is a 'good illustration of what makes the genre resonate with today's audiences'.[6] Released in the United States on 24 July 1998, the film had a production budget of US$65 million, was the highest grossing film of 1998 and at the time of writing has grossed some US$216.3 million in the US and US$480 million worldwide.[7] *SPR* is, therefore, an important entry point into examining how WWII has been written and understood in contemporary film.

Beyond its popularity at the box office, there are at least three reasons why *SPR* is a particularly important film for us to discuss in exploring some of the intertextuality between film and geopolitics. First, *SPR* was widely received and applauded for its 'realism' and for providing an 'authentic' historical document of the D-Day landings in particular and WWII in general. Numerous reviews praised the film for being 'grisly realistic',[8] the 'most authentic war movie',[9] and 'the most realistic depiction of war ever put on screen'.[10] It is, according to Cohen writing in *The Washington Post*, a film that 'obliterates the myth ... that death in combat is a dignified affair' and 'is far truer than ... the videotapes shown during the Gulf War ... narrated by Gen. H. Norman Schwarzkopf'.[11] *SPR* is interesting, therefore, because of its supposed authenticity and truthful account of WWII.

Second, because of *SPR's* perceived authenticity it is important to critically explore its particular narratives and images of WWII and their ideological force in the context of the late 1990s. The film opens with an unnamed veteran visiting the Normandy graveyards from where the viewer is then taken back to Omaha Beach and the D-Day landings in 1944. After 25 minutes of carnage we are returned to the home front and War Department officials discovering how, from a group of four enlisted brothers only one, Ryan (Matt Damon), remains alive, and the officials order him to be found and returned home safely. The central plot then follows Captain John Miller (Tom Hanks) and his squad as they embark on their mission to find Ryan. When found, Ryan predictably refuses to abandon his new brothers and urges Miller's squad to join them in doing something worthwhile for the war effort in defending their bridge position from an advancing German convoy. With the bridge successfully defended and Miller dying he urges Ryan to 'earn this' and the viewer is returned to the present to the Normandy graveyards with an elderly Ryan desperately asking his family 'Was I worth it?', 'Have I lived a good life?'. Throughout the film numerous questions are raised concerning sacrifice, duty, patriotism, virtue, warfare and responsibility and the way they are presented in the film speak very much to contemporary concerns and make *SPR* a product of the late 1990s and a 'landmark of *fin de siecle* American triumphalism'.[12]

Third, with WWII repeatedly providing a frame of reference grounding contemporary conflict in virtuous endeavour, in the final section of the essay we explore the intertexts between *Saving Private Ryan* and the 2003 Gulf conflict and ask how *SPR* forms part of a more general *moviesation* of geopolitical discourse. First, however, it is worth further exploring the role and functionality of film as a kind of modern-day geopolitician.

STAGING GEOPOLITICS: HOLLYWOOD AS GEOPOLITICIAN

As we watched the 2003 Gulf War unfold it was clear how Hollywood and the movies were central to its representation and staging. With the war, in part, a consequence of the terrorist attacks on 11 September 2001 it is also worth recalling how the language and images of the movies were central to our experience and understanding of those attacks. Immediately after 11 September numerous commentators noted Hollywood's apparent ability to predict the events in films like *Die Hard* (1988), *Executive Decision* (1996), *Air Force One* (1997), *The Siege* (1998) and *Armageddon* (1998) and in a related fashion the twinning of the attacks with a rhetoric of Pearl Harbour recalled the 2001 Hollywood blockbuster of the same name.[13] Moreover, in October and November 2001 Pentagon officials began a series of meetings with Hollywood directors, screenwriters, scenarists and specialists in disaster movies in order to 'solicit the help of Hollywood in the war against terrorism' and imagine possible future scenarios and responses to them.[14] This close relationship between Hollywood and the Pentagon continued into the 2003 Gulf conflict with the employment of top Hollywood set designers to assist in the design of the US media centre. Arguably this further underscores the importance of Hollywood to the arrangement and staging of vision and perspective and highlights the production of an important and influential 'logistics of perception'.[15]

The convergence of film, Hollywood and the representation of contemporary conflict raise important issues about the production of subject positions through cinema and the cinematic distancing of subjects from the realities of conflict. It also highlights some of the ways in which film is cast in the role of director on the world geopolitical stage, (re)writing the scripts of global politics as theatre. Of particular interest here is the way in which the Second World War is today packaged by the 'Hollywood war machine' as spectacle. In order to investigate the production of WWII in film and its connotative power in contemporary discourse it is useful to think about cinema and geopolitics in the terms of Mackinder's vision of the world as a 'stage' and international politics as theatre. Here, according to Ó Tuathail the subject position of the geopolitician is one of director arranging and interpreting the geopolitical stage with the 'mise-en-scene of the geopolitical gaze' being one of 'setting a stage scene or arranging a pictorial representation' – a '"setting into perspective" of the competing forces of current international politics'.[16] The geopolitician is a detached viewing subject surveying a worldwide stage and 'the relationship between the viewer and the worldwide stage is akin to that of a removed observer watching a theatrical production or a spectator viewing a panorama'.[17]

We can extend this analysis to argue that today film fulfills the role of geopolitician in producing a pictorial representation of international politics although the representational relationship between film and warfare is not new and has a long genealogy. While Mackinder was producing international politics as theatre, the theatrical stage in the turn-of-the-century music hall tradition was also staging productions of international politics.[18] Alongside the music hall, other popular vehicles of imperial vision such as the imperial exhibition, post cards, and panoramas were important in fulfilling 'a craving for visual representation of the world',[19] and the cinema and its imperial epics of the 1930s reinvigorated and continued the early traditions with their pictorial and visual production of the stage of international politics.

Emerging from these early traditions. therefore, film has, to a certain extent, always fulfilled the role of geopolitician. During WWII the British, American, and German film studios worked overtime narrating the war effort often diverting valuable military equipment and personnel from the real to the virtual war in order to assist production. In the early years of the Cold War film studios in the United States came under pressure from the House UnAmerican Activities Committee (HUAC) to produce a string of low-budget 'Reds under the beds' films that depicted, produced and reified the threat of domestic communism, and helped redefine Americanism and the boundaries of American identity. Similarly, in the latter stages of the Cold

War Jeffords argues that films like *Rambo: First Blood Part II* (1985) and *The Terminator* (1984) helped create the 'hard body' that served as a national emblem generating support for Reagan's remilitarised and tough foreign policy.[20]

In seeking to explore the implications of the cinematic production of geopolitical images and narratives on viewers of such films we have found the work of Virilio instructive. In *War and Cinema* Virilio[21] explores at length the relationship between film, perception and military technology arguing that they have developed alongside each other with advances in film technology changing the nature and experience of warfare for combatants. Pointing to the 'osmosis between industrialized warfare and cinema'[22] he suggests that 'alongside the "war machine" there has always existed an ocular ... "watching machine" capable of providing soldiers and particularly commanders, with a visual perspective on the military action underway'.[23] The result is a particular 'logistics of perception'[24] where a 'war of pictures and sounds is replacing the war of objects (projectiles and missiles)'[25] leading to a 'derealization of military engagement'.[26] Like Der Derian,[27] Virilio is primarily concerned with the effects and implications of virtualization on commanders and combatants but his arguments concerning cinematic derealization, falsification and dematerialization can equally apply to the production of an ethical distance between the movie going public and the realities of warfare.[28] Here, therefore, through an examination of *SPR*, we explore these aspects of film and geopolitics. In particular, we ask how WWII and combat was staged in *SPR* and, when according to Virilio, we have been plunged into an entirely 'cinematic vision of the world' where film and news feeds provide the 'raw material of vision', what mythology of WWII does *SPR* produce that can then be drawn upon in the staging of contemporary conflict.

GETTING TO THE TOP OF OMAHA BEACH TOGETHER: *SPR* AND THE OCULAR 'WATCHING MACHINE'

> It's easy to point out a couple of shots that are obviously very graphic, but it's the accumulation of the sequence on Omaha Beach that's supposed to help the audience understand the physical experience of combat. I didn't want to do something I've done with many of my other movies--allowing the audience to be spectators. Here I wanted to bring the audience onto the stage with me and demand them to be participants with those kids who had never seen combat before in real life, and get to the top of Omaha Beach together.[29]

One of the dominant themes to emerge from the reception of *SPR* at the box office was its supposed realism and authenticity. This is perhaps unsurprising when we consider the lengths Spielberg went to in his pursuit of realism. The Rankean entreaty to discover 'what actually happened', to seek emprical objectivity and 'proof' and to pursue truth about the past and historicity, runs throughout Speilberg's approach to *SPR*. In interviews Spielberg discussed how his actors were prepared in a military style boot camp and equipped with painstakingly researched and sourced authentic uniforms, webbing and weapons. After listening to veteran's recollections Spielberg littered the beaches with 1,000s of dead fish for the D-Day scenes and he even discussed how his choice of actors was governed by a search to find actors with 1940s-type faces:

> Kids in 1940 looked different. You can see it in the pictures from the war. They were closer to their immigrant roots. I went for faces that were like the faces I'd seen in the documentaries, like Eddie Burns ... He had the quintessential '40s face.[30]

Much debate about *SPR*'s authenticity centred on scenes of the Omaha beach landings that opened the film in a 25-minute avalanche of gruesome mayhem, disorientation and death. Spielberg shot these scenes with a hand-held camera with overexposed film stock where the frames jitter about in imitation of actual WWII footage and with blood splattered on to the lens the audience is made to feel the cameraman's fear of being caught in the crossfire. The intention was to draw the audience into the film, to take them to Omaha Beach, and let them experience warfare from a soldier's perspective. In filming these scenes Spielberg claims he 'decided to play the role of combat cameraman more than a director' and was 'trying to get as close as I could to the experience of what a combat soldier at Omaha Beach might experience'.[31] The result is certainly disorientation for the viewer and the casualness and randomness of death is heightened by the fact that audiences had not yet been introduced to a single character and lack an 'anchor' through which to view the unfolding carnage.[32] The sense of claustrophobia created here and the frenzied praying, vomiting and shaking (amongst the soldiers on the beach) further underline this impression.

With assistance from cinematographer Janusz Kaminski, Spielberg desaturated the colours of the film in order to emphasize the pale greens of the uniforms and landscapes, the blue-grays of the water and skies and the flesh tones of the characters, whilst the red of blood leaps out of this context for stark contrast. Interestingly the audience's own visual memories of events relating to WWII will have been largely drawn from newsreel footage of the times and so Spielberg finds his 'authenticity' by re-animating that footage in new 'modern' ways resulting in a kind of 'ersatz documentary' style.[33] Spielberg has also said that he took inspiration from Robert Capa's photos of Omaha beach:

> There was terror in each frame … Each frame was blurry, shaky and messed up, chaotic. They're frightening to look at, horrifically kinetic, and each photograph told the entire story of what it was like to be in combat. I thought, Well, if I could do something like that at twenty-four frames a second, that would be interesting.[34]

As combat photos and archive newsreel footage have become the iconic symbol of the D-Day landings since the end of the war, Spielberg conforms the images of his film to fit our visual knowledge of those events and our own frames of reference, which come to audiences partly through newsreel and documentary footage. The effect of this strategy in the film is powerful as the movie seeks to touch and align itself with the patriotic collective imagination.

The power of cinema for Shaviro is one of 'radical passivity'[35] wherein one has no choice but to look and see. The state of passivity is one of 'cinematic fascination'[36] with the viewers distracted attention held captive. Shaviro picks up on the basic tactility and viscerality of cinema that produces 'real effects *in* the viewer rather than merely representing phantasmic reflections *to* the viewer'.[37] Spielberg goes for the visceral in *SPR*, drawing the viewer onto Omaha Beach rather than *reflecting* the 'object' of the landings to spectators. The film, in this sense, is less a derealisation of combat and more an attempt at reification. Rather than provide a detached viewpoint from above Spielberg attempts to place the viewer on to the stage, to provoke reactions, and to allow viewers to feel and experience first hand the sense of horror and disbelief endured by 'citizen soldiers' at the time.

Not surprisingly, the Omaha Beach scenes and the effect they had on audiences dominated early reviews of the film. According to Matthews, the hand-held cameras shove us right into the middle of the violent maelstrom 'so close to the devastation we can hear bullets smashing bones, feel the splash of blood and saltwater against our face, and see the sweat and vomit of fear mixing with the smoke and dust of battle'.[38] This is a 'literal apocalypse now' but the effect on the shell-shocked viewer is perhaps 'nearly pornographic'.[39] This graphic verisimilitude at times overwhelmed audiences with many moved to tears by the film in general, and the opening sequences in particular. Jim Tharp, the distribution chief at Dreamworks, reported that 'the film seemed to have a very strong reaction on audiences … in some cases, people exiting the theatres were crying. In other cases, they could not initially talk'.[40] Similarly, the *New York Times* reported that 'some members of the audience leave in silence and burst into tears outside the theatre',[41] and according to *The Washington Post*, *SPR* is so 'unrelenting in its appalling honesty that few combat veterans will emerge from it without crying and trembling all over again'.[42] Moreover, in a preview screening Stephen E. Ambrose author of a collection of books about WWII and hired by Spielberg as a historical advisor, was so affected by the film's realism that he 'crawled under his seat and asked the projectionist to stop for a moment'.[43] He later reportedly argued that 100 years from now *SPR* will be *the* picture people watch when they wish to understand what WWII was about.[44] *SPR* is, therefore, becoming the ultimate frame of reference about WWII.

SPR's graphic realism was, therefore, central to producing the film's historical authenticity and establishing the truth of WWII. In this sense *SPR* fulfills the traditional role of the geopolitician objectively representing the realities and truths of international conflict. Its realism was, for many critics, what made *SPR* a new type of WWII film, but Bassinger argues that advances in technology and looser censorship restrictions allowed *SPR* to be more physically realistic in its depiction of warfare, but this does not mean *SPR* was more psychologically and emotionally realistic than earlier WWII films. Films such as *Bataan* (1943), *Beach Red* (1967) (which *SPR* clearly draws upon), *Cross of Iron* (1977) and *The Big Red One* (1980) were all, in their day, criticized for their grisly realism and violent nature.[45]

The supposed realism can also be seen as part of the wider Hollywood treatment of war by coming to represent 'the aesthicizing of violence, the sanitization of war'.[46] The deaths of US and German soldiers, moreover, are not treated equally. The heart-wrenching deaths in *SPR* are those of American soldiers but when German soldiers are shot down they go down like ten pins and *they stay down*. The film carefully constructs the Germans as different. In contrast to the youth and innocence of the Americans we find hardened professional soldiers whose faces are 'typically contorted by hatred for Americans'.[47] Nobody deserves this carnage we are told but clearly some don't deserve it more than others: 'their deaths [German soldiers] are *movie* deaths. And the more agonizingly the Americans suffer, the happier we are to see the German's slaughtered. The realism just ratchets up the enthusiasm.'[48]

Spielberg himself saw a close relationship between graphic violence and historical truth. On a promotional tour he 'warn[ed] parents and young people that this film may not be their cup of tea'.[49] The MPAA also added a stringent caveat to the U-14 rating it was given, noting that the film 'Includes intense, prolonged and *realistically graphic* sequences of war violence and language' (emphasis added). Nonetheless Spielberg says that had he been given a NC-17 rating 'I would have worn that like a Purple Heart, with pride and dignity, because I felt that the truth about what happened at Omaha Beach was long overdue'.[50] He also said he was determined that *SPR* would not be another 'slap in the face' to watching veterans: 'There have been 84 World War II films that showed something else. This would have been the 85th slap in the face to the men who died knowing the truth'.[51]

CINEMATIC CATHARSIS: THE SECOND WORLD WAR AS JUST, UNAMBIGUOUS AND HEROIC

> With *Saving Private Ryan*, war is good again.[52]

Although some critics initially argued that the gruesome opening scenes made *SPR* an anti-war film, the majority argued it upheld and reinforced the virtue of WWII as the 'good war'. The *New York Times*, for example, argued *SPR* showed WWII 'as a just and necessary war'[53] that 'restore[d] the nation's heroic image' by 'wiping clean the collective American conscience after the trauma of the Vietnam war'.[54] Critics on the right in the *Weekly Standard*, *The Washington Times* and *National Review* did complain about the lack of reflection on the 'higher purpose' of WWII but according to Krauthamer these criticisms are disabled by the simple fact that in the popular imagination 'World War II speaks for itself. It needs no spin. Only a moral idiot can doubt its justice.'[55]

But WWII does not speak for itself and it is spun in *SPR*. Structurally the war is spun as virtuous by drawing upon genre convention to sustain a narrative that places it within the mythological and heroic pattern of the WWII combat genre.[56] Thus we have the hero Captain Miller (Hanks), the group, and the classic difficult mission. There are hardly any women in the film,[57] and the opening scenes follow convention by declaring a 'true' time and date for what follows – again reminding audiences that this is a truthful and honest movie about an ultimately good war. The group in search of Ryan undertake an episodic journey across a combat landscape and obey genre conventions with their internal tensions, suspicion of outsiders, ingenuity in impossible circumstances, and the group itself is composed from classic WWII genre characters drawn from and illustrative of America's melting pot:

> the tough, battle-scarred sergeant (Tom Sizemore as Sgt. Horvath), the wisecracking Brooklynite (Edward Burns as Rieben), the 'Sgt.York' style godly Southern sharpshooter (Barry Pepper as Private Jackson), the feisty Jewish guy (Adam Goldberg as Pvt. Mellish), the salty Italian-American (Vin Diesel as Caparzo), the dedicated medic (Giovanni Ribisi as Wade) and the sensitive writer and outsider (Jeremy Davies as Cpl. Upham).[58]

The group, however, are not just 'a metaphor for American diversity and tolerance … they are the very essence of American civilization and its values'.[59] Private Ryan himself is a farm-boy from Iowa. Captain Miller, the uncommon common man, is the teacher-turned-soldier, slightly embittered and obsessive, self-enclosed yet tough. Many wartime films used the model of average guys plucked from Texas or Brooklyn 'who loved their everyday life in America or the girl next door'.[60] Miller's hand is seized up regularly with shakes and tremors, he cries, and through the vulnerability he portrays on film, we 'come to see clearly a decent man of the sort that America was theoretically meant to produce'.[61]

It is, moreover, very much *American* values and valor on display. *SPR* gives the impression that WWII was fought without allied help and also 'preserves the WWII image of American soldiers as inherently averse to bloodshed and cruelty'.[62] Some mistakes and bloodletting may occur in the fog of war but fundamentally the US soldier is different to the enemy and able to act with restraint in difficult circumstances. *SPR* thus constructs an image of the virtuous US soldier, but it also creates a virtuous state. WWII is rendered as a Manichean struggle between good and evil, and the United States clearly fighting with God on its side[63] demonstrated with sniper Private Jackson regularly praying before unleashing a fatal shot: 'Blessed be the Lord my strength, which teacheth my hands to war and my fingers to fight'.

The virtue of the US state is also upheld in that the central plot serves to demonstrate that the state 'has a heart'[64] and conjures a 'belief in the benevolence of authority'.[65] The soldiers fight out of patriotic duty, along with a desire to return home and resume domestic arrangements. Patriotic sacrifice is an important frame of reference in the film from the opening scenes in the graveyard of the Stars and Stripes. The film's action appeals to the *collective memory* of Americans about the heroes of the Second World War – their suffering and their triumph. As the film progresses, and the characters begin to develop, symbols of Americana totally pervade the work. The mother of James Ryan is informed of the death of her other three sons framed in the golden, autumnal colours of the Mid-West. Iconically, this sort of farming community symbolises all that is 'true' and good about America – it is earthy, honest, hard-working and God-fearing, the heartland of the United States. Additionally, the significance of Abraham Lincoln here is underlined as we hear twice his famous letter to Mrs Bixby, comforting a mother who lost five sons in battle. That invocation of Lincoln also reminds of the greatest American war, the Civil War, the cause of freedom it stood for, and Lincoln's greatest speech, the Gettysburg Address.[66]

Importantly, the film closes with a sequence set in the present. All the way through the film the question arises of how 'worthwhile' the mission is and it culminates with the issue of Private Ryan living a life worthy of those who died in WWII which for Basinger 'directly challenged the "me" generation sitting in the audience. (Are *you* doing anything worthwhile? People *died* for this country)' (emphasis in original).[67] The hero's dying admonition is 'Earn this' and *SPR* is all about showing Americans what they needed to earn. This question begs another, however, as Cohn and Weber argue in their reading of *SPR*: 'What ever justifies the taking of all those lives in a war?'[68] Landon similarly describes *SPR* as a 'morality play' in that it raised the question of what obligations war survivors and subsequent generations have to the soldiers who sacrificed their lives.[69] *SPR* claims that survivors of the war (like Ryan) and subsequent generations need to recognise their 'debt of honour' to America's brave combatants. The film is thus very much centred on the 'moral individual in heroism and in pain'[70] but it is this focus that makes *SPR* a product of the 1990s rather than 1940s or 1950s. According to Bodnar[71] films of the 1940s produced narratives of the 'people's war' emphasising themes like democracy and unity. *SPR*, in contrast, through Miller in particular, constructs the figure of the moral individual as the guardian of the future at the expense of democratic society: 'Past, present, and future are now contingent on standards of individual behaviour rather than on democratic ideals such as the quest for equality, a just capitalism, or citizen participation in political life.'[72]

With its emphasis on individualism, unilateralism, patriotism and duty *SPR* writes WWII in the context of *fin de siècle* domestic and international concerns and, as such, forms part of a larger Janus-faced WWII nostalgia industry. The style of *SPR* combines the intense violence and earthy dialogue of the post Vietnam *Apocalypse Now* era but its story is very much the product of the classic wartime epics of the 1940s and 1950s.[73] What is new in *SPR* is the 'post Vietnam project of consolidation' and the traces of what Derrida would call 'guilty nostalgia' for the war years where morality was seen as unambiguous.[74] As Baudrillard argues, 'when the real is no longer what it was ... nostalgia assumes its full meaning. There is a proliferation of myths of origins and of signs of reality.'[75] Post-1990 filmic productions of WWII like *SPR* appeal to nostalgic instincts and longings for a simpler bygone era of unambiguous glory and represent a desire to revisit and refight previous battles.[76] The WWII nostalgia regularly extends into political discourse. Der Derian discusses the Republican Convention and the wheeling out of WWII veterans on 'National Security Night'.[77] According to G.W. Bush,[78] this is 'the greatest generation[79] in history'.

SPR went some way towards painting an image for audiences of US WWII vets as leaving an extraordinary legacy and as representing the ultimate form of US citizenship. In the process, as Auster argues, 'WWII has become the indispensable symbol of American patriotic virtue and triumph. It can be brought forward to exalt American arms and the American spirit whenever contemporary events require it'.[80]

SECOND WORLD WAR/SECOND GULF WAR INTERTEXT: TWINNING RYAN WITH JESSIE

> In *Saving Private Ryan* you're not watching a bunch of battle-hardened John Wayne soldiers. You're watching a bunch of scared kids who just got out of high school.[81]

> They are calling it 'Saving Private Lynch'. Jessie Lynch is her name ... The story of her rescue is right out of a major motion picture: Blackhawks

> landing in the dark, a fire fight on the way in and out, a diversion across town by the Marines so they could slip in and get her out. The real story is almost more incredible than that.[82]

In seeking to explore the twinning of film on the Second World War and the Second Gulf War it is important to discuss how cinematic intertexts became central to reporting the rescue of Jessica Lynch in April 2003. Jessica was a nineteen year old PFC rescued by US special forces from an Iraqi hospital in Nasiriyah where she was being held prisoner. Lynch became instantly recognisable to many across the country as the first US prisoner to be rescued from behind enemy lines since the Second World War.[83] The episode came as a welcome and reassuring reminder to Americans that its forces could strike effectively – anywhere and any time – in the pursuit of rescue. It was a major opportunity for patriotic pride in both Jessica's supposed bravery and the dedication of her rescuers, and many observers noted the symbolic importance of her rescue in turning the mood towards the war. The rescue mission was seen to prove that the resolve of US forces was as strong as ever and represented a kind of 'made-for-television catharsis'.[84] As Katz has pointed out, 'she and her rescue absolutely had an impact on media coverage, morale of the troops, and even the stock market'.[85]

Of particular importance here are the many cinematic intertexts in the reporting of this 'heroic' rescue mission behind enemy lines and its moviesation by journalists. The headline across most of the US (and indeed British) media was either 'Saving Private Jessie', 'Saving Private Lynch', or 'Saving Jessie Lynch' – each one making a direct reference to *Saving Private Ryan*. In doing so *SPR* provides a direct frame of reference for understanding the whole rescue and the rescue itself becomes conjured as another WWII combat genre film. The hero, the group, the difficult mission and the just war are all connoted in such headlines.

Interestingly, the rescue was filmed by a combat cameraman accompanying the assault team and the images were seen around the world the next day. Like the Omaha Beach scenes in *SPR* the viewer is taken to the action and given a soldier's eye view of combat. The Omaha Beach scenes in *SPR* strangely precede and mimic those of an embedded reporter and, according to Hodgkins,[86] should be seen in the context of the live video feeds and television images shown in the 1991 Gulf War. Now, in 2003, we have similar scenes and similar parallels. Through the reporting of Jessica's salvation (and like the narratives of *SPR*) we come to witness a kind of journalism reminiscent of the 1991 conflict 'whose goal is to highlight the heroism of the United States as we engage in a battle whose successful resolution is never in doubt from the moment the opening credits … roll across the screen'.[87] Thus, the embedded journalists were themselves part of a moviesation of the Second Gulf War, 'Spielbergising' the production of the conflict (although they did not generate combat footage of any note), taking viewers through the equivalent of Omaha Beach.

The differences with *SPR* are also apparent in that there are no deaths in this film – *Saving Private Jessica* and by implication twenty-first century warfare has all the positive connotations of WWII combat genre and none of the negative ones.[88] In this sense it is possible to argue that *SPR was* a historical document – contemporary warfare is just and virtuous, but also different to the chaos of Omaha Beach, and Jessica's rescue mission/movie proves it. In the current Hollywood vogue for remakes the saving of Private Lynch is an *SPR* remake – updated and presented as an advert for today's warfare. Its central plot and indeed central message is the same. According to White House Press Secretary Ali Feischer the operation had 'demonstrated the heart and soul of the US military',[89] although many later criticized it as, in Captain Miller's words, 'a PR mission'.

While *Saving Private Ryan* provided the most common cinematic reference point to representing Jessica Lynch's rescue it is important to note that it was not the only cinematic reference. Representations of the rescue were, in fact, suffused with film intertexts on the part of the media reporting the events *and* the participants recounting the events. In the case of the former, journalists situated Jessica's story within the captivity narrative[90] with Jessica cast as 'one of those damsels in distress of the silent movie era, like Mary Pickford tied to the traintracks'[91] or as a kind of latter-day Fay Wray 'struggling in the paws of King Kong'.[92] Lynch's supposed heroics in resisting her capture also saw her referred to as a 'teenage Rambo'.[93] The press also reported on the Hollywood interest in the story and of book and film agents fighting for the exclusive as the 'as-yet-untold story instantly hardened into a Hollywood – style fable of heroism'.[94] According to media critic and author Jon Katz the story of this 19-year-old heroine was 'like a Hollywood script, almost too much'[95] and similarly Morse noted how perhaps even Hollywood 'could not have dreamed up a more singular tale'.[96] Thus not only was the reporting of Jessica going missing in action and then into captivity framed by reference to Hollywood movies but also the very nature of the rescue was pure Hollywood. Not surprisingly A&E put out a special presentation *Saving Jessica Lynch* only two weeks after the rescue.

As noted above it was not only the news media who reported the rescue in the language of Hollywood. US Colonel Harren Warren also saw the events as Hollywood, stating 'it worked perfectly. It was like *Black Hawk Down*, except nothing went wrong'.[97] In a similar fashion Dr Anmar Uday, an Iraqi doctor at the hospital where Jessie was held captive, remarked on similarities with Hollywood film but also pointed to the possibility that the whole rescue *was* a movie.

> It was like a Hollywood film. They cried, 'go, go, go' with guns and blanks and the sound of explosions. They made a show – an action

> movie like Sylvester Stallone or Jackie Chan, with jumping and shouting, breaking down the doors.[98]

The idea that Jessica's rescue was pure Hollywood comes as part of a barrage of criticisms and unanswered questions concerning discrepancies between initial Pentagon accounts and those of eyewitnesses. Early reports recounted the Pentagon script of Lynch's heroic fight in the desert where she sustained bullet and stab wounds, her possible torture and interrogation, and the daring rescue mission to rescue her from Iraqi armed guards. The BBC's *Correspondent* programme, however, later reported that she had no stab or bullet wounds, had received the best medical treatment available including blood donated by hospital staff, and the 'rescue' was from a hospital long deserted by the Iraqi army. The film of *Saving Private Jessica*, it was suggested, included US forces firing blanks to liven up the action.[99] Surely then Virilio[100] is right to suggest that in warfare today the target area has become a 'cinema location, the battlefield a film set out of bounds to civilians'.

CONCLUSIONS

> 'Am I a good man?' and 'Have I led a good life?'

> If Tom Hanks can die senselessly then anybody can and *nobody will* (emphasis added).[101]

> 'Real' politics is increasingly played out by virtuous means: we seek to find in Hollywood .... what we can no longer find in Washington.[102]

For some commentators, 11 September 2001 signaled the end of certain trends in the production of combat films, ushering in a 'new phase of consolidation for the Hollywood War Machine'.[103] Films (and the language and images of films) have been key reference points in making sense of and representing the 2003 Gulf War and indeed the September 2001 attacks on America that precipitated the recent war in Iraq. Movies figure in the popular political imagination as ways of imagining and enframing global political change. While a relationship between war and films is not unique to this conflict the ubiquitous nature of war/movie intertexts does suggest that the availability of discourses enabling war to be experienced as a movie represents an intensification of the *moviesation of war discourse*. In this essay we have examined in further detail the intertextual relationship between movies and the Second Gulf War. In terms of the moral grammar of war, cinema allows this intertext to create the illusion of virtue and virtuosity. In particular we have sought to illustrate something of the 'theatricalisation' of geopolitics through cinema, drawing out the dematerialisation and derealisation of war in certain kinds of films. In so doing, we want to flag up the provision through cinema of a kind of *ocular watching machine* for experiencing mediated war from afar. The production of international geopolitics as filmic theatre is important here as is the playing out of politics by 'virtuous means' as Der Derian writes in talking about the search for something in Hollywood that is no longer on offer in Washington. The new phase of consolidation that Pollard refers to still involves a heavy militarism, violence and patriotism and will continue to involve occasional convergences with 'real-world combat':

> the venerable Hollywood war machine is sure to persist in creating heavily militaristic, violent, patriotic films designed to satisfy American viewers' seemingly insatiable appetite for dramatic, bloody and vicarious war experiences that, from time to time, will probably converge with real-world combat.[104]

SPR was often seen as 'gritty', 'brutal' or 'vital' and its director was arguably obsessed with a kind of Rankean realism. As Cohen[105] notes, the film was often seen as balancing up the truth, revalidating particular virtues: 'D-Day was, as we all know, a smashing strategic success – but now we have an inkling of the horrors that accompany even lopsided victories'.[106] Is it not, however, meaningless to write of 'realism'[107] in war movies when most of us have little idea what war really looks or feels like?[108] To police fact from fiction here is meaningless and also likely to fail in that it can also 'turn a blind eye to what popular representations can tell us about politics and the politics of desire bound up in *interpretations* of historical events'[109] (emphasis in original).

We began this paper with the question of how WWII films produce narratives of virtuous war. In prioritizing the narration of international politics as filmic theatre we are also keen not to neglect the role of the viewing subject and the question of their positioning in the constructed global theatre of geopolitics. What, then, are the messages of the film – perhaps they are open-ended (itself a departure in films directed by Spielberg)? Is one message that it's acceptable to kill a POW? Is another that American intellectuals like Colonel Upham somehow condoned the Holocaust 'through their sympathies with the Germans as civilized human beings'?[110] What is of paramount importance here is that in *SPR* Captain Miller's dying words make it possible for the Second World War generation to be turned into the 'embodiment of American ideals of self-sacrifice for the twentieth and twenty-first centuries.[111] Virtue is embodied in the sacrifice of Captain Miller such that the Second World War becomes for Americans that 'mythic, edenic moment when the entire nation bent itself to victory over evil and barbarism'.[112] *SPR* is thus a serious inspiration and reference point in the twenty-first century from which to proclaim American triumphalism. The frames of reference constructed in *SPR* thus serve to represent a powerful and seductive blend of triumphalist mythologies including:

> the religious aura of America's martyred sixteenth president, Hollywood's classic World War II films, the ahistorical refusal to acknowledge the major role that the allies played in the downfall of fascism, the celebration of the American intellectual tradition, and the glorification of American GI's as the world greatest fighting men.[113]

This became an attractive way of narrating a successful war fought by honest soldiers and good kids, especially those fresh out of high school. The priority given in *SPR* to the need for future generations to 'earn it', to earn and justify the sacrifices of previous combatants through our actions in the historical present can also be read in other ways however. Crucial to the narrative of *SPR* is the question of what comes next, what comes after or follows such brave sacrifices. Are audiences of the film expected to retain a debt of honour and gratitude to *everyone* who has died and suffered in WWII? Should viewers and readers of 'Saving Private Jessie', therefore, also honour the Iraqi soldiers and civilians who died in the war? Does the injunction to earn it also include a debt to the British and American troops still dying in Iraq months after the end of the conflict?[114] Does our commitment to Iraq include a long-term commitment of generations – a difference between an 'ethics of intention' and an 'ethics of responsibility'.[115] As one church leader noted in twining the salvation of privates Ryan and Lynch, their deaths invite our response:

> When we confront the cross we are deeply moved. 'Saving Private Ryan' and 'Saving Jessica Lynch' makes us think of Jesus dying to save us. The death of Jesus invites our response. If it is true, have I earned it? Has your life been worth it.[116]

At the end of the film the old Mr Ryan asks his wife whether he is a good man. Uncomprehending, she nonetheless reassures him. The implication of this is to ask are we worthy of Lincoln, our Founders and those who 'gave the last full measure of devotion'? Do we deserve to be linked over the generations with these heroes? These are the questions the film poses to us at its end. With the words of Abraham Lincoln echoing in the background (pride, sacrifice, glory, freedom — we're all dying sweetly *pro patria!*), the weepy vet returns to the screen.

The action involved in the saving of Private Ryan symbolizes what national salvation requires: We must be good. We must be good, because among the survivors is one of the most contemptible soldiers ever portrayed in a movie—the intellectual Corporal Upham, an interpreter of French and German who does not know common American slang. Presuming intellectual and moral superiority over his fellows, this base coward is in fact their inferior. He is the prototype of the Vietnam War protester—an arrogant moraliser in speech who in deed collaborates in a brutal slaughter. His fatuous questions to his wife: 'Have I lived a good life?', 'Am I a good man?' are placed as a reminder that Americans must remain good and righteous out of respect for the sacrifices of people like Miller. Indeed, Miller's final words to Ryan: 'Earn this' suggests the necessity for the United States to unite and strive to secure America, lest Miller die in vain. Therefore, since Ryan must live as the future of America, there needs to be a martyr who acts as a lesson to those who reap the benefits of such sacrifice. Miller therefore serves this function, and as the character we are encouraged to foster allegiance with, his death will ultimately have the most impact, reinforcing the dominant vision of the heroic *patria*.

Spielberg opted for the more clear-cut moral universe of WWII over the messy complexity of the Vietnam conflicts. *SPR* is an attempt to forget Vietnam as MacCabe[117] has noted. The film is more than just an attempt at forgetting uncomfortable events of the past, however. Spielberg's story of GIs risking their lives hunting down an endangered dogface to save his life is itself a kind of measure of what it was that Americans lost between the Second World War and Vietnam.[118] Arguably this kind of virtue in virtuous war is re-celebrated in the Second Gulf War through the media frenzy surrounding Jessica Lynch. One weakness that has been noted of Spielberg before is that he has 'a compulsion to pasteurize even the mud of battle', steering away from the complex and contradictory.[119] Thus *SPR* opens a number of wounds but does not necessarily investigate them *SPR* is slick in its production but this has a price in that the film is 'softened by an almost neurotic slickness that keeps getting in the way of the important, explosive issues it [*SPR*] raises'.[120] *SPR* is therefore all about the triumph of the American spirit – so much so that all the films triumphs are seemingly won without allies. The United States' increasing willingness to act unilaterally in world affairs since the end of the Cold War and certainly post-11 September 2001 demonstrates that, like many of the narratives embedded in SPR, this too provides a frame of reference for contemporary concerns.

NOTES

1. T. Pollard, 'The Hollywood War Machine', *New Political Science* 24/1 (2002) p.133.
2. G. Ó Tuathail, 'The Effacement of Place: US Foreign Policy and the Spatiality of the Gulf Crisis', *Antipode* 25 (1993) pp.4–31
3. G. Ó Tuathail, *Critical Geopolitics: The Politics of Writing Global Space*, (London: Routledge 1996).
4. J. Der Derian, *Virtuous War: Mapping the Military-Industrial-Media-Entertainment Network* (Oxford: Westview Press 2001) p.xv.
5. Ibid. p.201

6. D. Sterritt, 'War Stories: From Patriotic Flag-waving in the 40s to the Re-release of "Apocalypse Now", War Movies still Resonate with American Audiences', *The Christian Science Monitor* (2001). <http://www.csmonitor.com/2001/0810/p13s1-almo.html>, accessed 14 April 2003.

7. The Numbers, 'Saving Private Ryan: Box-office and Production Data', 2003 <http://www.the-numbers.com/>, accessed 19 August 2003.

8. J. Mathews, 'The Good War, in Dying Color', *Newsday,* Part II/Weekend (1998) p.BO3.

9. J. Verniere, 'Movie Review: Guts and Glory; Brilliant, yet Flawed, "Saving Private Ryan" will Shake you to your Core', *The Boston Herald*, 24 July 1998, SCE p.503.

10. T. Lawson, 'Tall Soldiers Action-packed and Profoundly Moving: *Saving Private Ryan* Salutes American Gallantry and Ponders Deeper Questions', *Detroit Free Press* Features (1998) p.1D.

11. R. Cohen, 'A Reminder from Speilberg', *The Washington Post*, 28 July 1998, p.A15.

12. A. Auster, 'Saving Private Ryan and American Triumphalism', *Journal of Popular Film and Television* 30/2 (2002) p.100.

13. C. Weber, 'Flying Planes can be Dangerous', *Millennium: Journal of International Studies* 31/1 (2002) pp.129–47.

14. S. Zizeck, *Welcome to the Desert of The Real*(London: Verso 2002) p.16.

15. P. Virilio, *War and Cinema* (London: Verso: 1989) p.1.

16. Ó Tuathail (note 3) p.30.

17. Ibid. p.29.

18. J. Mackenzie, *Propaganda and Empire: The Manipulation of British Public Opinion, 1880–1960* (Manchester: Manchester University Press 1986).

19. Ibid. p.16.

20. S. Jeffords, *Hard Bodies: Hollywood Masculinity in the Reagan Era*, (New Brunswick, NJ: Rutgers University Press 1993). Film has not only supported hegemonic geopolitical representations but has also served as a medium of resistance in films like *Dr Strangelove* (1964), *Platoon* (1986), *Full Metal Jacket* (1987) and *Three Kings* (1999).

21. Virilio (note 15).

22. Ibid. p.58.

23. Ibid. p.3.

24. Ibid. p.1.

25. Ibid. p.4.

26. Ibid. p.1.

27. J. Der Derian, 'Simulation: The Highest Stage of Capitalism?', in D. Kellner (ed.), *Buadrillard: A Critical Reader* (Oxford, Blackwells 1994) pp.189–209.

28. From its very beginning in the nineteenth century and its development as a medium alongside trade fairs 'cinema became the major site for a trade in dematerialization'; Virilio (note 15) p.32). It is the ability of films to superimpose themselves on to geostrategies that is important here: 'when *Around the World in 80 Minutes* shone in lights outside of cinemas in the 1930s, it was already clear that film was superimposing itself on a geostrategy which for a century or more had been leading to the direct substitution, and thus sooner or later the disintegration, of things and places'; Virilio (note 4) p.47. Before the advent of cinema panoramas had been popular, particularly in some of the great imperial exhibitions of the early nineteenth century.

29. .Spielberg, cited in R. Ebert, *Chicago Sun*, 19 July 1998.

30. Spielberg, cited in S. Hunter '"Private Ryan": Steven Spielberg's Unflinching Tribute to the Men who Conquered Hell', *The Washington Post* Style (Friday, July 24, 1998) p.B01.

31. Spielberg, cited in R. Lyman, 'True to the Timeless Fact That War is Hell', *New York Times*, 19 July 1998, p.11.

32. See C. Cohn and C. Weber, 'Missions, Men and Masculinities: Carol Cohn Discusses Saving Private Ryan with Cynthia Weber', *International Feminist Journal of Politics* 1/3 (1999) pp.460–75. As Cohn and Weber put it, 'one of the things that's striking about the *Ryan* war scene is that it was not shown through anyone's eyes' (p.472).

33. Hunter, 'Private Ryan' (note 30) p.BO1.

34. Spielberg, cited in (Entertainment Tonight/July 21, 1998) <http://www.mattdamon.com/extra/audio/ET-spr2.ram/>.

35. S. Shaviro, *The Cinematic Body* (Minneapolis, MN: University of Minnesota Press 1993) p.48.

36. P. Molloy, 'Theatrical Release: Catharsis and Spectacle in *Welcome to Sarajevo*', *Alternatives* 25 (2000) pp.75–90.

37. Cited in Molloy (note 37) p.84.

38. Mathews (note 8) p.BO3.

39. Ibid.

40. B. Weinraub, '"Ryan" Lands with Impact in Theaters across US', *The New York Times*, 27 July 1998, p.E1.

41. B. Weinraub, 'At the Movies; Ryan Headed for an Oscar?', *The New York Times*, 24 July 1998, p.E14.

42. P. Fussell, 'The Guts not the Glory of Fighting the "Good War"', *The Washington Post*, 26 July 1998, p.C01.

43. K. Gabbard, 'Saving Private Ryan's Surplus Repression: Film Review Essay', *International Journal of Psychoanalysis* 81(2000) p.177.

44. D. Ryan, 'Appraising the Value of Life in a Sea of Blood: *Saving Private Ryan* is Virtuosic', *The Philadelphia Enquirer* Features Weekend (July 24, 1998) p.3.

45. J. Bassinger, *The World War II Combat Film: Anatomy of a Genre* (Middletown, CT: Wesleyan University Press 2003).

46. Der Derian, *Virtuous War* (note 4) p.166.

47. Pollard (note 1) p.129.

48. L. Menand, 'Jerry Don't Surf: Review of Saving Private Ryan', *New York Review of Books* 45/14 (24 September 1998) p7.

49. J.I. Brennan, 'Filmmaker Launches a Tough Defense of the R-rated "Saving Private Ryan", which Contains Grisly Images of War', *The Los Angeles Times*, 15 July 1998, p.1.

50. Spielberg, cited in ibid. p.1.

51. Ibid.

52. V. Canby, 'War Movies: The Horror and Honor of a Good War', *New York Times*, 10 August 1998, p.E1.

53. E. Rothstein, 'Rescuing the War Hero from 1990s Skepticism', *The New York Times*, 3 August 1998 p.E2

54. Canby (note 52) p.E1.

55. C. Krauthammer 'Debating *Pvt. Ryan*', *The Washington Post*, 14 August 1998, p.A25.

56. Basinger (note 45); Pollard (note 1).

57. Women only appear in the cemetery scenes and 'back-home' in Iowa. Indeed, the film's tag-line 'The mission is a man' has been the subject of some debate by feminist writers such as Cynthia Weber and Carol Cohn who have highlighted its 'purging of the feminine'; Cohn and Weber (note 33) p.461.

58. M. Wilmington, 'Apocalypse then: Steven Spielberg's Violent "Saving Private Ryan" Depicts both the Horror and Humanity of World War II', *Chicago Tribune*, 24 July 1998, p.2.

59. Auster (note 12) p.102.

60. J. Bodnar (2001) '*Saving Private Ryan* and Postwar Memory in America', *The American Historical Review*106/3 (2001) p.805.

61. Miller's character is also the most idealised of old American soldiers in *SPR*, with a paternal imago, filling in for the father of the missing Private Ryan. The captain fights not because of a love of war and fighting but because of the time-tested wish of every American GI to return home to his wife and family. For Gabbard there are thus many 'Oedipal issues' underlying Spielberg's work here. Corporal Upham (with whom Spielberg was said to most identify) comes to symbolise the large numbers of baby boomers who sat out the war and thus refused to be tested in the crucible of battle. See K. Gabbard, 'Film Review Essay. Saving Private Ryan's Surplus Repression', *International. Journal of Psychoanalysis* 81 (2000) pp.177–9; see also T. McCarthy, 'Combat Viewed through Visceral Cinema', *Variety Magazine*, 20 July 1998, p. 45.

62. Bodnar (note 60) p.805.

63. J. Hodgkins. 'In the Wake of Desert Storm: A Consideration of Modern World War II Films', *Journal of Popular Film and Television* 30/2 (2002) pp.74–84.

64. Canby (note 52) p.E1.

65. S. Hunter, 'Strafing "Private Ryan"', *The Washington Post*, 9 August 1998, p.G1.

66. See also 'Saving "Private Ryan" from the Conservatives', Ken Masugi copy available at <http://www.leaderu.com/humanities/masugi.html/>.

67. Basinger (note 45) p.262.

68. Cohn and C. Weber (note 32) p.469.

69. P. Landon, 'Realism, Genre and *Saving Private Ryan*', *Film and History* 28/3–4 (1998) pp.58–63.

70. Bodnar (note 60) p.807.

71. Ibid.

72. Ibid. p.817.

73. Wilmington (note 58); Basinger (note 45).

74. Gabbard (note 61) p.178.

75. Der Derian, *Virtuous War* (note 4) p.156.

76. This might also include Spielberg's fascination with the Second World War, which he has spent a good deal of his career trying to make sense of. Like many of the 1960s boomers he grew up watching *Combat* (1962–1967), the hit TV show directed by Robert Altman about a squad of soldiers fighting in Europe during the Second World War; Verniere (note 9), which was itself one of a number of popular sources of this notion of a 'good war'. As a 14-year old budding film-maker he directed a movie (cast with friends and neighbours) called 'Escape to Nowhere' about US–German battles in the Second World War. He has since returned to the war obsessively with two Indiana Jones films (1981, 1989), with *Empire of the Sun* (1987), with *Schindler's List* (1993), the comical *1941* (1979) and *Always* (1989) as well as *SPR* (1998). He added yet another wartime film for the HBO television network in 2001 with the *SPR* spin-off series *Band of Brothers*. His current projects include a film version of Arthur Golden's best-selling *Memoirs of a Geisha* and again returns to the Second World War era. Greenberg even argues that this is Spielberg's way of working through his relationship with his father; H.R. Greenberg, 'Dulce Et Decorum: A Review of Saving Private Ryan", 1991, copy available at <http://www.isidore-of-seville.com/d-day/7.html>, accessed 3 December 2003.

77. Der Derian, *Virtuous War* (note4).

78. 28 May 2001. President George W. Bush presents medals to 21 Navajo Code Talkers during a ceremony honoring their military service at the US Capitol, 26 July 2001 (*Windtalkers* released June 2002).

79. Speaking on Veterans Day, 6 November 2002, Bush said of veterans that 'these patriots have inspired our nation with their courage, compassion and dedication ... securing the blessings of liberty' and called on all US citizens to remember these sacrifices and to rededicate themselves to today's challenges of securing an enduring peace; G.W. Bush, 'By the President of the United States of America: A Proclamation', Veterans Day press release, 2002, <http://www.whitehouse.gov>.

80. Auster (note 12) p.104.

81. Spielberg, cited in S. Stark, 'A Close Encounter with Spielberg: Director shows his Mastery with WWII Epic', *Detroit News* Film Reviews, July 25, 1998, p.1.

82. B. Williams, NBC News, 2 April 2003.

83. J. Adler, 'Jessica's Liberation', *Newsweek*, 14 April 2003, p.42.

84. A. Stanley, 'A Nation at War: The TV Watch', *The New York Times*, 18 April 2003, p.B9.

85. J. Katz, cited in J. Thomas, 'Rescued POW to Face a Bigger Enemy', *Milwaukee Journal Sentinel*, 9 April 2003, p.D1, copy available at <http://www.jsonline.com/enter/books/apr03/131967.asp?format=print/>.

86. J. Hodgkins, 'In the Wake of Desert Storm: A Consideration of Modern World War II Films', *Journal of Popular Film and Television* 30/2 (2002) pp.74–84.

87. Ibid. p.76.

88. The celebratory tone and spectacle of the whole Jessica Lynch episode can be criticized for the downplaying of the 11 US servicemen and women who died in the events that lead to Lynch's capture by the Iraqis.

89. Ali Fleischer, cited in T. Wilkinson and G. Miller, 'War with Iraq/Midnight Rescue', *Los Angeles Times*, 3 April 2003, p.1.

90. During the first Gulf War, the first American woman prisoner, Specialist Melissa Rathbun-Nealy, was reportedly treated like a celebrity by Iraqi captors: They compared her bravery to Sylvester Stallone's and her beauty to Brooke Shield's, and fed her so well she actually gained weight in captivity; Adler (note 83) p.42. Soon after the First Gulf War had finished, numerous agencies packaged the narrative into videos and CD-ROMs for our viewing at home in case we missed it first time round.

91. J. Neuman, 'Lynch Now Network's Objective', *Los Angeles Times*, 22 June 2003, p.22.

92. Stanley (note 84).

93. Ibid.

94. Ibid.

95. Katz, cited in J. Thomas, 'Rescued POW to Face a Bigger Enemy in Media', *Boston Globe*, 8 April 2003, available at <http://www.jsonline.com/enter/books/apr03/131967.asp/>, accessed 3 December 2004.

96. J. Morse, 'Saving Private Jessica', *Time Magazine*, 14 April 2003, p.66.

97. Harry Warren cited in J. Morse, 'Saving Private Jessica,' *Time Magazine*, April 14, 2003, p.66.

98. Cited in J. Kampfner, 'The Truth about Jessica', *The Guardian*, 15 May 2003, G2, p.2.

99. Ibid.

100. Virilio (note 15) p.11.

101. C. MacCabe, 'Bayonets in Paradise', *Sight and Sound*, February 1999, p.14.

102. Der Derian, *Virtuous War* (note 4) p.158.

103. Pollard (note 1) p.138.

104. Ibid. p.139.

105. E. Cohen, 'What Combat does to Man: *Private Ryan* and its Critics', *The National Interest*, Winter (1998) pp.82–8.

106. Ibid. p.88.

107. In a review of critics responses Cohen discusses the opening sequences as reality as whether they match up to reality or not; and overwhelmingly they do. Cohen, for example, argues that on a second viewing the opening scenes reveal 'a particular improvisation and determination that, in the end, made the landings a success. A thread of tactical skill, only intermittently visible to those watching, underlies the wiping out of a German bunker'; ibid. p.87.

108. J. Wrathall, 'On the Beach', *Sight and Sound* 8/12 (1998) pp.34–5.

109. Weber (note 13).

110. Wrathall (note 108) p.35.

111. Auster (note 12).

112. Ibid. p.104.

113. Ibid.

114. Some 37 US troops were killed in Iraq in the month of November alone and more US troops have now been killed since 1 May 2003 than died in the hostilities during the invasion of Iraq itself. Some 19 British troops have died since the end of hostilities on 1 May 2003, bringing the total number of British servicemen that have died in Iraq to 52 (BBC News, 10 November, 'US Warns of more Iraq attacks', <http://www.bbc.co.uk/news/>).

115. H. Shinoda, 'The Politics of Legitimacy in International Relations: A Critical Examination of NATO's intervention in Kosovo', *Alternatives* 25 (2000) pp.515–36.

116. Easter Sunday Sunrise Address, Sydney Opera House, Reverend Dr Gordon Moyes, Wesley Mission 2002.

117. MacCabe (note 101) p.14.

118. T. Appelo, 'Iraq War: The Movie', *Seattle Weekly*, 2 April 2003, p.23.

119. M. O'Sullivan, 'Spielberg wins Battle, not War', *The Washington Post*, 24 July 1998, p.N37.

120. Ibid.

Screening Geopolitics: James Bond and the Early Cold War films (1962–1967)

KLAUS DODDS
Department of Geography, Royal Holloway, University of London, Surrey, UK

INTRODUCTION

> In my experience all Prime Ministers love intelligence because it is a secret weapon they have. The intelligence report used to arrive in special little boxes and it gave them a belief that they had a direct line to something that no ordinary departments have. They never did without it and they never minded spending money on intelligence in my experience.[1]

> The conflict is melodramatic and Manichean. Bond ... represent[s] goodness without ambiguity, and they face a series of supremely sinister, evil, and grotesque villains, whose unbridled lusts and cruelty drives them towards world domination.[2]

Over the last ten years, disciplines such as International Relations (IR) have embraced the so-called 'return of culture' in a variety of ways. It has been raised not only as an issue of forthcoming insecurity in the form of civilisations and corresponding clashes but also in the context of how popular culture might influence/shape/reproduce the foreign and security policies of governments.[3] Moreover, Cynthia Weber has recently noted, 'accessing visual culture through popular films allows us to consider connections between IR theory and our everyday lives. Using popular films in this way helps us get a sense of the everyday connections between 'the popular' and 'the political'.[4] In the case of political geography and more specifically geopolitics, however, the notion of a 'return' is perhaps misplaced.[5] The introduction of the term 'popular geopolitics' (linked as we shall see below to the critical geopolitics corpus) signified a renewed interest in the manner in which popular ideas about global political space help to reinforce or resist the foreign policy discourses and practices of political elites. New research into film, cartons, television and popular journals such as the *Reader's Digest* arguably complemented a longer-standing interest in political geography's relationship to popular culture and visual media (as opposed to music and sound geographies for example).[6]

Notwithstanding geography's long-standing interest in film and the representation of places on celluloid, it is only comparatively recently that academic sub-fields such as geopolitics (whether of the critical variety or not) have shown any interest in visual media with the exception of maps.[7] This is, as Michael Heffernan has noted, an extraordinary lacuna because geopolitics, understood as an intellectual practice dating from the turn of the twentieth century, has always been highly visual.[8] However, with the introduction of a self-consciously critical geopolitical literature, a new interest arose in other forms of visual media and their relationship to elite and popular cultures. Political geographers such as John Agnew, Joanne Sharp, Gerard Toal and Simon Dalby helped to expand the intellectual field of geopolitics by reminding us that the practices and representations of global political space do not reside exclusively in politically elite cultures. In other words, ideas and images associated with world politics 'leak' into popular culture and may either reinforce and or contest dominant political understandings such as the depiction of the Soviet Union as a threatening other in Cold War America.

This expanding literature of critical geopolitics spawned an inter-locking area of interest labelled 'popular geopolitics'.[9] Over the last decade, theoretical sophistication has been combined with empirical diversity and important interventions such as Joanne Sharp's detailed analysis of the *Reader's Digest* during the Cold War have been realised.[10] Likewise, new analyses of the print news media have appeared which demonstrate how popular geopolitical perspectives needs to drawn upon and engage with related literature in media and cultural studies. Arguably, popular geopolitical analyses of film have not evolved in quite the same manner in terms of textual and or methodological sophistication. In our eagerness to explore how particular representations of places and people reinforce and or contest particular foreign and security policy discourse, we (those who write and or engage with the critical geopolitics literature) have been insufficiently attentive to the nature and practices associated with film.[11]

Rather than seek to analyse film conventions and or the film industries, Hollywood being just one aspect of this global enterprise, the object of this essay is to extend a geographical appreciation of film and in particular the James Bond series. As noted earlier, film (especially those produced in Hollywood, or Pinewood Studios in the case of James Bond) often dramatises contentious issues and highlights value conflicts. In so doing, geographical locations and or geographical framing are rarely if ever passive backdrops to these acts of dramatisation. The portrayal of people in space, the locales (sometimes fictional) used in films and the role of movie frames help to define space and human relationships therein.[12] The depiction of places also assist in the formulation of narrative structures and thus, in the context of the James Bond films, the movement of Britain's premier spy is an important and persistent feature of his intelligence missions.

These odysseys help to highlight two features – first, Bond's travel plans help to convey a sense of glamour and mystique to his missions. He either glides across international borders with the minimum of fuss under a false identity and occupation and or endures an enforced intimacy with his adversaries such as on board the personal jet plane of Auric Goldfinger. It is salutary to note that for viewers in the 1950s and 1960s, before the invention of mass airline travel and package holidays, Bond's missions took him to places that very few would have been able to visit (see Fig. 1). The line between tourism and international spying may be a fine one. As Michael Denning has noted, 'it stands, for Bond, as an idyllic paradise, as a more authentic culture, and as a source of threat and betrayal ... Bond is given a more privileged access than the average tourist. He is taken by his secret work, into secret worlds'.[13] Second, Bond's escapades across space help to generate a sense of eventual climax, which usually witnessed a titanic – often highly individualised – struggle involving him and his main adversary. Intimate and confined places, irrespective of environment such as the personal jet of Auric Goldfinger and or the compartments of the *Orient Express*, are important in producing a narrative ending.

As part of initiating a richer geopolitical appreciation of film, one could consider how the mapping of films such as the James Bond series might facilitate such a task. As Moretti has shown with his analysis of the European Novel, such an exercise assists in asking how space and locations are used to fashion the narrative structure including the beginnings and endings of particular films.[14] So rather than simply noting that post-Cold War American action films have developed different story lines (following the collapse of the Soviet Union), we actually investigate how places are embedded into the movie structure and the original screenplay. This might entail a closer reading of films and, as Lisa Benton helpfully concludes, 'film opens a window on the sphere of human relations, depicting how people move through space, how they interact with objects in the landscape, how they interact with space ... How they interact with each other in this space'.[15]

The aim of this essay is to further contribute to a geographical engagement with one particular visual media – film, and more specifically the feature films involving the British secret agent, James Bond. Given an earlier essay on *From Russia with Love* (1963), I am neither proposing to rehearse the basic bio-geographical details of Ian Fleming[16] nor aiming to replicate the important contributions by individuals such as Tony Bennett, James Chapman, Jeremy Black, Andrew Lycett and David Cannadine.[17] The second section nonetheless aims to explore why a focus on James Bond might be profitable with regard to the continuing evolution of popular geopolitics. Thereafter, the conversion from the Fleming novels to the films are scrutinised with a particular focus on how and why the plots and geographical locations were altered by personnel attached to Eon Productions and United Artists.

FIGURE 1 Mapping James Bond, 1962–1965

The role of Richard Maibaum, the chief screenwriter for 13 Bond films, deserves special attention because of his role in translating the James Bond novels to the screen. For the sake of manageability, attention is devoted to the first five films – *Dr No* (1962), *From Russia with Love* (1963), *Goldfinger* (1964), *Thunderball* (1965) and *You Only Live Twice* (1967). All five were huge hits with the viewing public in North America, Europe and elsewhere such as Japan. Finally, the essay concludes with how this interest in the James Bond film series might have wider implications for how we approach the development of popular geopolitics.

WHY JAMES BOND?

Notwithstanding my own continued enjoyment of the James Bond films,[18] it behoves me to explain why these cinematic productions might assist us in developing an understanding of film and popular geopolitics. Four basic reasons might be offered at this stage. First, the James Bond film series (dating from 1962 onwards) have been extremely popular not only in Europe and North America but also in Asia and Africa. It is routinely estimated by James Bond scholars that at least a quarter of the world's population has seen at least one Bond film.[19] Thus, whatever one might critically note about the storylines and plots, a great number of people have been exposed to the exploits of Commander James Bond/007. This is not of course to claim that all the viewers enjoyed the experience and or consumed the films in a similar manner.[20] Second, it is possible to derive some interesting parallels with inter-state relations and the practices associated with intelligence gathering. In other words, the films self-consciously mimic (and of course exaggerate in places) the diplomatic and intelligence intercourse of the post-1945 world. The creator of James Bond, Ian Fleming, was a former wartime intelligence officer and had access to an extraordinary range of people, including the Kennedy brothers.[21] The films are filled with explicit references to so-called allies and friends (e.g. the United States) and adversaries (e.g. the Soviet Union and China). Third, as a consequence of its commercial success, the longevity of the James Bond series allows us to consider how the plots and story lines have been able to account for global political shifts from the Cold War to the post-Cold War era, for example. Thus, the plot line attached to the latest James Bond film, *Die Another Day* (2002), combined terrorism, so-called 'rogue states' such as North Korea and the figure of the 'evil genius' in a manner which would have been readily appreciated by viewers (especially Americans in the aftermath of 11 September 11 and George Bush's 'War on Terror') across the world. Fourth, the James Bond story lines have used the figure of the 'evil genius' (as we shall see later) to complicate understandings of inter-state relations. In other words, James Bond's missions frequently involve him confronting a dangerous figure that is, at best, only loosely connected to Cold War adversaries such as China and the Soviet Union.

The introduction of Commander James Bond, Britain's premier spy, therefore, remains an extraordinary moment in the cultural history of the Cold War.[22] Despite the claims to the contrary from the film producers, Cubby Broccoli and Harry Saltzman, the James Bond adventures were always sensitive and sensitised by the prevailing Cold War conflict. This was perhaps not surprising given the claim by the historian Tony Shaw that, 'virtually everything, from sport to ballet to comic books and space travel, assumed political significance and hence potentially could be deployed as a weapon both to shape opinion at home and to subvert societies abroad'.[23] Even if the evidence thus far is fragmentary, we know that the James Bond novels and films were immensely popular with the very president who weathered the 1962 Cuban Missile Crisis – John F. Kennedy – and his brother Attorney General Robert Kennedy. Indeed, according to one biographer of Robert Kennedy, 'JFK was such an aficionado of Fleming's novels that, by enthusing about them, he helped make the British thriller writer [Ian Fleming] a top best-seller in the US'.[24] Shortly after its official US release, President Kennedy arranged for a private showing of the film *Dr No* in the White House.[25]

Over the last 40 years, the Bond phenomenon has attracted a coterie of critics and commentators eager to analyse and interpret the symbolic, cultural, political and sexual significance of the fictional super-spy.[26] The figure of James Bond has been interpreted as a sop to British imperial decline,[27] a rear-guard response to the rise of feminism,[28] an example of fragile post-war masculinity,[29] an illustration of a longer literary British spy tradition[30] and a retrogressive colonial fantasy.[31] Arguably one of the reasons for the enduring appeal of Bond is the connection to the geopolitics of the Cold War and the use of place-based imagery to convey intrigue, mystery and danger. Often interpreted by trade press reviewers as simply 'exotic locations', the places depicted in the James Bond films are more than simply backdrops to the unfolding spying mission. They actively contribute to the geo-graphing of the Cold War as a series of Manichean struggles between good and evil. The main adversary is often an evil genius figure rather than a representative of an opposing superpower, however. In particular the earliest Bond films starring the Scottish actor, Sean Connery, explicitly and implicitly utilise a series of assumptions about place in order to convey a sense of danger filled with individuals frequently differentiated on the basis of race and ethnicity such as Dr No and Auric Goldfinger. While the films demonstrate that there can be no 'safe' places in the midst of a Cold War, they also paradoxically illustrate that 'security' can only be secured via the sterling endeavours of James Bond rather than a vast array of conventional and nuclear weapons of the NATO alliance.

FROM THE NOVEL TO THE SCREEN: NARRATING JAMES BOND

Ian Fleming died at the relative young age of 56 in August 1964 and thus was only able to witness the filming and subsequent release of three films, *Dr No* (1962), *From Russia with Love* (1963) and more fleetingly *Goldfinger* (1964). Apart from being interested in the filming process, Fleming also accrued a tax advantage from being on location. If the producers employed him then he did not have to pay tax in the UK, provided his services were performed overseas. He might have seen the making and release of the film *Thunderball* (1965) had it not been embroiled in a lengthy legal dispute involving Kevin McClory and the ownership of the film script. While the dispute was eventually settled out of court in December 1963, the translation of the novels into film involved negotiations with Cubby Brocolli and Harry Saltzman of Eon Productions and the Hollywood film company, United Artists. As Tino Balio notes in his historical account of United Artists, 'the James Bond films are quintessential examples of products tailored for the international market. Financed by an American major partly with British film subsidy funds, produced by two expatriates who had incorporated in Switzerland, and based on a popular series of espionage novels that played off Cold War tensions, the James Bond films were shot in exotic locales featuring a cast of mixed nationalities that was headed by a star of universal appeal'.[32]

In order to understand more fully the translation from the James Bond novel to the film, it is necessary to explore how United Artists and Eon Productions shaped the Bond ethos and in particular the embedded importance of formulaic story telling. Harry Saltzman, a Canadian born producer, contacted Fleming after reading the aforementioned *Life* magazine story featuring Kennedy's favourite novels. In the midst of another bout of poor health and eager to create a trust fund for his son Casper, Fleming sold Saltzman an option on the Bond books. The deal was not sealed until Broccoli met with Fleming and finalised a deal which meant that exclusive movie picture rights were secured for all the published James Bond novels with the exception of *Casino Royale*.[33] In return, Fleming was to receive $100,000 per book and 2.5 per cent of the net profits from each film. As Balio explains, the deal was premised on the notion that there would be a substantial series of films because Saltzman was required to option a new book for a film every 18 months or else the rights to the novels would revert to Fleming. If the series succeed, and this was still not at all certain given the modest sales of Fleming novels in the United States, then Saltzman was empowered to claim the motion picture rights to the character of James Bond.

Armed with the film option, Broccoli and Saltzman approached United Artists and a deal was eventually struck in April 1962, even if the first film budget was to be a modest one – $140,000 for property and screenplay, $40,000 for the director and $140,000 for the entire film cast. In order to qualify for a British government-sponsored Eady Plan subsidy, it was also agreed that the cast would have to be predominantly British, including the lead actor, and shot on 'British' locations which could include the British Empire and/or Commonwealth. While Broccoli and Saltzman wanted to produce *Thunderball*, impending legal action meant that *Dr No* was chosen as the first 'official' Bond film. Real-life events helped to make *Dr No* an attractive option as American rocket testing at Cape Canaveral was bedevilled by claims that some of the rockets were going dangerously astray. Thus the basic plot involving an evil genius interfering with the telemetry of American rockets seemed remarkably prescient. Given that the location of Fleming's novel was the island of Crab Key, the filming process was centred close to Fleming's home (Goldeneye) in Jamaica. As a member of the British Commonwealth, Jamaica qualified for an Eady Plan subsidy.

Richard Maibaum, who had worked with Broccoli on an earlier Antarctic-based film *Hell Below Zero* (1954), was appointed the chief screenplay writer. After a wartime career working in the US Armed Forces' Combat Films Division, Maibaum and a co-worker Alan Ladd were responsible for a series of post-war productions including an espionage film *O.S.S.* (1946) and *Paratrooper* (1954).[34] Before his association with the James Bond series, Maibaum was associated with British and American productions as well as a stint as an executive producer with MGM-TV between 1958–1960. Alan Ladd recommended him to the producers Cubby Broccoli and Irving Allen of Warwick Films in the mid 1950s. This relationship with Broccoli later embraced Harry Saltzman, when Eon Productions secured the film rights to Ian Fleming's novels. Maibaum was charged with turning the novels into films and as such he helped to transform locations, dialogue and story lines sometimes in a dramatic fashion. As Adrian Turner notes, 'Maibaum must be regarded as a key contributor to the Bond ethos and the Bond machine; its creation surely owes nearly as much to him as it does to Ian Fleming, or to Saltzman and Broccoli, or to Sean Connery and Ken Adam'.[35] While he was paid $40,000 by United Artists to transform the novel *Dr No* into a movie script, the initial offering was not welcomed by United Artists' Bud Ornstein, who complained that 'I must tell you I have not been impressed with Maibaum's work and only hope he will come up with something much better this time'.[36] Despite these initial misgivings, Maibaum remained the chief screenwriter for 13 of the Bond films and retired after the production of *Licence to Kill* (1989).

With the benefit of the Richard Maibaum's papers, including his screenplays and commentaries on the figure of James Bond, we are better able to see where he intervened to transform Fleming's novels. This is significant because some films such as *From Russia with Love* (1963) worked closely with the novel while others, such as *You Only Live Twice* (1967), substantially differed from the original story line, which did not feature a space-based intrigue involving the two superpowers. Maibaum, inspired by the stories of Alexander Dumas (such as *The Three Musketeers*), recognised that these transformations resided within a generic Bond story:

> Basically, in a way, the pattern of the story is the same in all the Bonds, and I think that is one of the attractions of the pictures. You have that engine working, the James Bond syndrome – with all the conspicuous consumption, the luxurious locales, the beautiful women, the larger-than-life villains. We've carried it much further than Fleming. Bond is absolutely dedicated and devoted to serving Queen and country; he never questions what he does or the morality of it [cf. *Licence to Kill* 1989, however] ... The Dumas books had a helluva [sic] influence on me.[37]

The first James Bond film, *Dr No*, provides interesting insights into how the basic geopolitical elements of the plot were changed from the original novel published in 1957 by Jonathan Cape. Moreover, it also reveals the importance of place in generating intrigue and danger. In the novel, Bond eventually discovers that the Chinese–German Dr Julius No has bought Crab Key and is intent (under the cover of running a guano-processing operation) on interfering with the telemetry of American rockets for sinister reasons. Confronted by Bond, Dr No admits in a rather febrile manner that:

> You are right Mister Bond. That is just what I am, a maniac. All the greatest men are maniacs. They are possessed by a mania, which drives them forward towards their goal ... I am, as you correctly say, a maniac – a maniac, Mister Bond, with a mania for power ... Mister Bond, power is sovereignty, Clausewitz's first principle was to have a secure base. From there one proceeds to freedom of action. Together, that is sovereignty. I have secured these things and much besides ... Who in the world has the power of life and death or death over his people? Now that Stalin is dead, can you name any many except myself? And how do I possess that power, that sovereignty? Through privacy.[38]

This austere vision of power was supplemented by a cunning plan involving the interference of the American rocket-testing centre on Turks Island. Dr No claimed to Bond that most of the testing failures such as the SNARK rocket were the responsibility of his facility on Crab Key. According to his devious planning, Dr No boasted to Bond:

> Their rockets would go mad. They would land on Havana, on Kingston. They would turn round and home on Miami ... There would be panic, a public outcry. The experiments would have to cease. The Turks Island base would have to close down. And how much would Russia pay for that to happen?[39]

If Dr No succeeded, America's scientific and technological advantage would be turned against itself and the British colonial dependency of Crab Key would ultimately pose a far greater danger than Communist Cuba. The United States' closest Cold War partner unwittingly made Dr No's operations possible when it sold Crab Key to him in 1943. While Bond later kills Dr No, it is clear that his operations were loosely connected to Chinese criminal networks. He is, in the main, a solitary evil genius.

In the first and second edition of the *Dr No* film treatment, Maibaum and his co-writer Wolf Mankowitz change the details and emphases of the original plot. In the opening scene, for example, the British secret service operative Strangeways is murdered by a gang of Chinese–Afro-Caribbean assassins. While the film retained that scene, it did feature another proposed corresponding scene involving the assassination of an American secret agent called Christopher on board the SS *Orinoco*. Christopher was killed just as he was trying to contact Strangeways to warn him that the ship (bearing explosives) was headed towards Crab Key. Bond is despatched to Jamaica in an attempt to discover who has murdered Strangeways. As he begins his investigations, he encountered an anti-Castro Cuban called Gomez who is eager to make Bond's acquaintance. As the film treatment progresses, it becomes clear that Gomez is actually a Cuban agent seeking to obtain arms and explosives from Dr No. As the film treatment noted:

> The picture is clearer now, Bond tells Leiter. Dr No is obviously supplying arms through Gomez to pro-Cuban elements in the Caribbean, and his guana [sic] operation is only a cover for his armaments company.[40]

The purpose of this fiendish scheming becomes clearer when Bond discovers that the crew attached to the SS *Orinoco* (flying a Cuban flag) intends to destroy a part of the Panama Canal and as a consequence create the impression that Cuba was responsible for this act of 'terrorism'. When he finally confronts Dr No with the 'Cuban' plot, Bond is informed that the conspiracy is actually more sinister:

> When the Panama Canal is blown up by a ship flying the Cuban flag it will draw undoubtedly reprisals from the American government against Cuba. The ensuing Russian counter-action should produce a highly desirable state of chaos, in which a resolute and highly efficient organisation like the Black Monkey [a criminal society] should become the dominant factor in the Caribbean.[41]

In a final twist to the plot, one of Strangeway's British–Jamaican companions, Hugh Buckfield, is revealed to be Dr No because, as Bond notes, both men have an identical Capuchin monkey as a pet! Dr No perishes in a gigantic fireball as Bond destroys the explosive-laden SS *Orinoco*.

The second edition of the film treatment dropped the idea of the monkey (disliked by United Artists) and Dr No assassinates Buckfield rather than assume his identity. While the rogue Cuban agent Gomez remained at the heart of the secret plan, the role of 'Red China' was emphasised rather than

Dr No and the Chinese criminal society, the Black Monkey. This was an important difference because the threat was explicitly territorialised. Dr No was no longer represented as just a Chinese–German evil genius with vague links to Chinese underworld. In an expanding second edition of the film treatment, he was described as a 'Red Chinese agent' who sought to convince Bond that he should join his organisation because:

> The destruction of the Gatun Locks of the Panama Canal will undoubtedly evoke reprisals by the United States against Cuba. This in turn will prompt Russian action against the United States. Such a situation is considered desirable by Red China. In the ensuing conflict there will be many areas where chaotic conditions will prevail. One of them will be the Caribbean. Dr No's activities will exploit these conditions for Red China's benefit and his own. He will more than ever need an efficient organisation ... In his opinion Bond would be an ideal replacement [for Buckfield].[42]

Despite the changes to Dr No and his political loyalties, the Panama Canal remained the centre of the drama rather than the novel's preoccupation with the interference of American rocket testing in the Caribbean. In the aftermath of the Suez Crisis (1956) and the struggle for control of the Suez Canal, the choice of the Panama Canal (opened in 1914) was opportune in the sense that it was considered to be a strategic passageway between the Caribbean and the Pacific Ocean. Moreover, as early as the 1950s, American strategic planners were worried that left-wing subversives (supported either by Cuba and or the Soviet Union) might seek to disrupt American and Western commercial traffic through an attack on the canal itself. In so doing, Maibaum's proposed film treatment shifted the centre of the intrigue onto an American-controlled Panama Canal Zone and thus it was easier to justify the enhanced presence of the CIA's Felix Leiter. With the help of Leiter, Bond precipitated an enormous explosion, which consumed Dr No in an explosive finale.

A committee composed of the producers Broccoli and Saltzman, the production designer Ken Adam and the director Terence Young worked and re-worked the script with due attention given to special effects and set production. Maibaum's private papers reveal how frustrating this collective editorial process could be:

> After I write and submit my first draft of the screenplay, revisions are discussed in committee ... The writer is both the anvil and one of the sledges. He participates as a clanger in this nerve-wracking anvil chorus while simultaneously recording it. The wear and tear on the psyche is understandingly harrowing.[43]

Maibaum's film treatment had been altered again on successive occasions so that by the time of the film's release, it was far closer to the original novel's plot. The references to the Cuban agent Gomez were removed, as was the explicit political link between Dr No and 'Red China'. Following Broccoli and Saltzman's caveat about minimising explicit geopolitical references to Bond's enemies, Dr No was revealed to have been a member of the non-aligned terror network, SPECTRE. When asked by Bond about his political backers, Dr No informs him that 'East' and 'West' are merely 'points on the compass' and have no significance for the trans-global ambitions of SPECTRE. The prevailing geopolitics of the Cold War have no ideological significance in so far as the demands made of the United States government reveal no particular geopolitical bias.

The other substantial difference between Maibaum's earliest film treatments revolves around the role of Bond's CIA counterpart Felix Leiter. Maibaum envisaged the operation to locate and destroy Dr No's activities as a joint Anglo-American mission because Leiter would also be intent on avenging the death of the US secret agent, Christopher. In the film, Leiter's role was modest and he does not, for example, accompany Bond on his covert trip to Crab Key. With the absence of the 'Christopher' dimension, Bond was able to dominate proceedings and Britain is spared the indignity (in the post-Suez Crisis era) of the Americans taking over an operation involving one of their own colonial territories. But this uneasy balance involving Anglo-American operations persisted during the filming process. In a memorable letter by Ann Fleming to Evelyn Waugh in February 1962, she noted that:

> On location with the film company producing Dr No: they were shooting a beach scene, the hero and heroine cowering behind a ridge of sand to escape death from a machine gun mounted on a deep sea fishing craft borrowed from a neighbouring hotel and manned by communist Negroes. The sand ridge was planted with French letters full of explosive – by magic mechanism they blew up the sand in little puffs. The machine gun gave mild pops but I was assured it will be improved on the sound track; all this endeavour was wasted because unluckily a detachment of the American Navy entered the bay in speed launches and buggered it all up.[44]

In his detailed review of 'The Sixties', Arthur Marwick identified eight defining features – the growth of sub-cultures and movements, individualism, youth, technology, international cultural exchange, consumer culture, civil rights and permissiveness especially with reference to sexual relations.[45] For many viewers and trade press reviewers, these earliest Bond films such as *Dr No* and *From Russia with Love* helped define the Cold War zeitgeist. As the *Film Daily* declared, Bond, 'played with a winning mixture of urbanity and masculinity by Sean Connery, is dispatched to Jamaica to investigate. Sean Connery learns that the murders were only part of a sinister plot for world domination executed by Dr No, a gifted, partly Chinese scientist.'[46] Or as *Kine Weekly* opined, 'it tells how James Bond, an immaculate, resourceful and fearless British agent, prevents an oriental master criminal, operating from a Caribbean island fortress, gaining world domination through atomic power'.[47] *Dr No* was premiered at the Pavilion Theatre in October 1962 and was released simultaneously in 198 theatres throughout the British Isles. It grossed a staggering $840,000 in the first two weeks and was extremely popular in other European countries including Italy, Germany and France. In May 1963, the film was released in New York and American audiences helped to generate a $2 million domestic and $44 million foreign gross.

After its release in April 1964, the second Bond film, *From Russia with Love*, subsequently grossed $12.5 million. It was the first film to be released during the Christmas season in the United States, a pattern that was to be followed by *Thunderball* when it was released on 21 December 1965. Although an extraordinary increase, the third film *Goldfinger* (1964) propelled the Bond film series into the major league of cinema box office receipts. As with earlier films, the film treatment of *Goldfinger* prepared by Maibaum and the production team departed from Fleming's novel and in this case the link with 'Red China' becomes an important part of the film's geopolitical rationale. Premiered in London in late September 1964, one month after the death of Ian Fleming, *Goldfinger* eventually grossed over $46 million worldwide and spawned a veritable industry of 'Bond/007' products including toiletries, shoes, model cars and the serialisation of the Fleming novels in *Playboy* magazine.[48]

REEL GEOPOLITICS: MAPPING JAMES BOND

Jeremy Black's *The Politics of James Bond* is an important intervention in the ensuing discussion involving the geopolitics of the Cold War and how it relates to the James Bond films of the 1960s.[49] While 'Bond is a figure designed to resist the decline of empire', the condition of Britain and the empire was fragile as the winds of post-colonial change swept across Africa, Asia and the Caribbean. By the time *Dr No* was released in October 1962 to the British viewing public, Jamaica was no longer a colony.[50] While many have noted that the James Bond films differ markedly from Fleming's novels in terms of the depiction of the adversary, less attention has been given to how, for example, the mode of travel and places assisted in the generation of geographies of evil and danger so necessary for the Bond plot lines.

While the role of the United States and the CIA is at times completely marginal or even absent (for example *From Russia with Love*), the Bond–Leiter relationship was fundamentally shaped by the British agent's superior intelligence gathering and analytical abilities. Bond is usually grateful for the resources of the United States but at same time mindful of the value of British pluck and ingenuity.[51] This 'brains–brawn' partnership was to be memorably depicted in *You Only Live Twice* (1967) in two dimensions. First, in complete contrast to Fleming's earlier novel, Bond tells Miss Moneypenny that he has no need for a 'beginners guide to Japanese' because he took a First in Oriental Languages at Cambridge.[52] Felix Leiter does not feature in the film as instead the Japanese Secret Service and its commandos provide essential support for the mission. In the novel, Bond 'had never been east of Hong Kong' and had no predilection for Far Eastern languages such as Japanese, let alone Chinese or Korean.[53] Indeed, prior to his mission he remained puzzled by his selection. The depiction of Bond in the film merely consolidated a view of a man possessed with extraordinary attributes and talents – a cunning linguist, as Miss Moneypenny memorably recorded in a later Bond film, *The World is Not Enough* (1997). As the reviewer for the trade press journal *Variety* noted, 'his well documented reputation as a bon vivant is here extended to his having placed a first in Oriental Languages at Cambridge and knowing that saki should correctly be served at 98.4° F. Audiences obviously relish these superhuman qualities the stuff of which mass daydreams and UA box office millions are made of'.[54]

Second, in the aftermath of the disappearance of American and Soviet spacecraft, a secret meeting was convened among the British, Americans and Soviets. After exchanging accusations and counter-accusations, the British official is shown to be contradicting his American counterpart when he dismisses the claim that the Soviets must be responsible for the disappearance of the American craft. Instead, the British official claims that their intelligence from the Singapore tracking station thought that the missing spacecraft might be found in an area close to the Sea of Japan. A range of places, nonetheless, helps to accommodate and extenuate the differences between Britain and the United States as depicted within the James Bond films.

Notwithstanding the fact that Singapore (independent in 1965 from Malaysia) was no longer a British colony, the inference regarding superior British intelligence appears laughable given the scale and extent of American resources, including its space-based programme. While the films as with the novels depict the British rather than the Americans on the frontline of the Cold War, they are able to function through discreet references to Britain's foremost contribution to Anglo-American intelligence operations – islands and enclaves. Places such as Hong Kong (Chinese intelligence gathering), Cyprus (Middle Eastern and Soviet intelligence gathering) and British Indian Ocean Territory (US military base on Diego Garcia) demonstrated the value of a scattered imperial dominion. Even an uninhabited outcrop such as Rockall in the North Atlantic was annexed by the Royal Navy in September 1955 because it was situated within the radio-electrical range of a test site for Britain's first nuclear weapon, the American-made 'Corporal'. Rockall, like Dr No's Crab Key, was a strategic vantage point from which to monitor the telemetry of the Corporal missile.[55]

Even if the British lacked the financial and political will to establish substantial forces 'East of Suez', the availability of real estate was significant at a time when the Cold War had become truly global in scope and intensity. As the CIA's Frank Wisner memorably confided to the British spy, Kim Philby, in the early 1950s: 'Whenever we want to subvert any place, we find the British own an island within easy reach.'[56]

The location of the Caribbean in the films *Dr No* (1962) and *Thunderball* (1965) is clearly significant precisely because it was a region where British imperial influence remained considerable, given that island colonies such as Jamaica, the Bahamas and Barbados were only to secure independence in the 1960s. Fleming's Caribbean-based novels are filled with reflections on colonial society and his misgivings about the loss of colonial influence in the region. However, the aforementioned films in general convey a sense of imperial continuity rather than dwelling on Bond's periodic outbursts of colonial angst. British officials in the form of the police and the coastguard continue to patrol the land and water around those British colonies. As Christopher Hitchens noted, it was one of those regions where the United States had not yet replaced Britain as the great power presence.[57] In actual fact, the United States, in the form of the 1961 Bay of Pigs fiasco involving an attempted coup against Castro's Cuban regime, had received a humiliating blow. While this was clearly not equivalent to the 1956 Suez debacle, it did point to the fact that even a superpower could suffer the occasional reversal in the field.

The struggle against communism lies at the heart of the Bond novels and there is less cultural nuance compared to the writings of John Le Carré and Graham Greene. When he takes to the field, Bond represents not only Britain but also the 'Free World'. However, these confrontations with communism are cross-fertilised by the introduction of evil geniuses and their criminal networks, usually supported or encouraged by China or the Soviet Union. The introduction of the multi-national terror/criminal network SPECTRE was arguably one of the most significant interventions in the making of the James Bond films. Fleming used the novel *Thunderball* (1961) to move the emphasis away from the Soviet spy service SMERSH. Despite the declining significance of Anglo-Soviet competition, the novels nonetheless provide reassurance that Britain retains (and must retain) a global presence.

This shift of political gravity allowed the producers to situate their earliest films within the confines of the Cold War while simultaneously enjoying an artistic licence to depict a variety of evil geniuses located across the world. In so doing two visions of political space are developed – a world dominated by territorial nation-states in the midst of a Cold War and second a global system infiltrated by a trans-boundary network of criminals using nation-states merely as bases for their activities. Remote islands and or subterranean headquarters were favoured precisely because it was assumed that the surveillance capacity of states would be unable to cope with these discrete locations. With the exception of *Thunderball* (1965), audiences were largely kept either unaware of the extent of SPECTRE's criminal activities and or its headquarters.[58] Early on in the film *Thunderball*, however, the audience was able to follow the movements of Largo, Number 2 in the SPECTRE organisation and thereby discovering that SPECTRE's (temporary?) headquarters were located in Paris within a building fronted by the International Brotherhood for the Assistance of Stateless Persons.

This seems particularly apt given SPECTRE's trans-boundary criminal activities and willingness to recruit operatives from East or West and North or South. Before embarking on a discussion of their 'most audacious' plan to steal two nuclear bombs from a NATO aircraft, Number 1 orders a formal review of criminal activity. The assembled SPECTRE operatives provide an 'area report' that includes details on the blackmail of a Japanese double agent, the assassination of a French defector to the Soviet Union (paid for by the French government), a £250,000 consultation fee for the Great Train Robbery in Britain and the distribution of Chinese-based drugs to the United States. Unhappy with the claim that only £2.3 million has been raised from the distribution activities of drugs, Number 1 ruthlessly electrocutes one of the operatives. After the summary execution, Number 2 explains the plan to demand a ransom of $280 million for the return of the two nuclear bombs, or else American and/or British cities will be threatened. Strikingly, the only government mentioned as being involved in any kind of financial and or political relationship with SPECTRE is France, a country later to withdraw from NATO's command structure in 1966. As a result, France was deeply distrusted by successive generations of Anglo-American governments.[59] Once again, places within the James Bond screenplays perform an important role in generating different conceptions of danger and intrigue.

The use of SPECTRE as part of the adversarial formula of shadowy networks and evil geniuses also relied on developing a series of racial geographies of danger. In keeping with all Fleming's Bond novels, the adversaries are never described as 'white' and or 'Anglo-Saxon'. Dr No was a Chinese–German scientist, Blofeld was a Greek–Polish engineering graduate and Auric Goldfinger was described as Central European and Jewish. Clearly these kinds of depictions hark back to monsters such as the Transylvanian Count Dracula and the evil Chinese doctor, Dr Fu Manchu, which filled the pages of Victorian novels.[60] Evil and race were welded onto one another in an attempt to convey how Western civilisation was in danger of being imperilled. The action adventure story (and subsequent translations into film) provided a ready format for such racial depictions of evil. The use of dark clothing and dark glasses in the Bond films merely compounds those simplistic depictions of evil along with the black-coloured claw hands of evil Dr No.

Places such as Turkey, featuring in the film *From Russia with Love* (1963), were never simply passive backdrops and or 'exotic locations'. The escape of Bond on the appropriately named Orient Express actively created a sense of an ideological and geographical boundary between an unsafe and claustrophobic Balkan space and a civilised and friendly European space (in this case Venice, Italy). Unusually for a Bond film, the movement of the train was traced along a map of South East Europe precisely to generate a sense of intrigue as Bond escapes from the atavism of the Balkans and its peoples. But it also enables a dangerous intimacy to flourish as Bond finds himself trapped on board the train with a ruthless assassin masquerading as a British secret agent. While Istanbul featured in earlier novels such as Graham Greene's *Stamboul Train* (1934), the location and the mode of travel within the film help to generate a collective sense of double dealing, sinister plotting and intrigue.[61]

In contrast, as soon as he stepped outside the confines of British airspace,[62] his world becomes more culturally complex and imperial spaces such as the Caribbean are depicted as spaces filled with dangerous and often 'childlike natives'. The opening scene of *Dr No* (1962) is perhaps the most explicit example, as three blind black men make their way through Kingston accompanied by a children's nursery rhyme ('Three Blind Mice'). As the film develops, the scene switches to a group of English colonials playing cards at an exclusive club where the only black men to be seen are employed as servants. As the British Secret Service representative Strangeways (who part of the Bridge playing foursome) leaves to make contact with his superiors in London, he is shot by one of the 'blind men'. Strangeways's death and subsequent exposure as a British agent (the blind men later murder his secretary and steal secret files) is then juxtaposed with a rare image of the calm efficiency of the British Secret Service's headquarters in central London. While chaos reigns in the colony, order rules supreme in the imperial centre.

When Bond arrives in Jamaica, he is confronted with a colonial locale in which many of the natives are barely capable of comprehending either sophisticated Western technology or Bond's cunning scheming. One of the most interesting relationships involves Bond and the Cayman Islander, Quarrel, who, although essential to the eventual success of the mission to destroy Dr No, cannot distinguish a vehicle armed with a flame-thrower from a dragon. This is a man who is haunted by superstition and tradition, as opposed to Bond's analytical style. Quarrel is never shown how to use the Geiger counter; instead, it is Bond who has at his disposal science and technology. Quarrel sails his boat by memory rather than with the help of a nautical chart. Moreover, Quarrel drinks rum to calm his nerves while Bond appears unflappable. Jamaica and its inhabitants provide an ideal setting for Bond's imperial heroics and as such highlights a reluctance to acknowledge a diminished role for a Britain entering perhaps unwillingly into a post-colonial world.[63]

Cold War espionage by its very nature tended to be highly covert and obscure places often played an important role in generating climaxes to the Bond films. The films of the 1960s actively reflected the kinds of subterranean worlds familiar to those engaged in secret operations against the Soviets and their Eastern European partners. Bond's subterranean movements were absolutely in keeping with real-life missions such as the 1955 Anglo-American Operation Gold/Stopwatch, which featured the construction of a secret tunnel under East Berlin for the purpose of collecting intelligence.[64] Despite his penchant for air travel, expensive hotels and fine wine, Bond spends a great deal of his time negotiating tunnels, secret passageways and underground bases such as the Crab Key covert control centre in *Dr No* (1962), the basement of Goldfinger's house (1964), the underground waterways of Istanbul in *From Russia with Love* (1963), and the spectacular volcanic base of SPECTRE in *You Only Live Twice* (1967).

These austere or confined locations perform at least four visual functions – first, they provide the geographical settings (and help define ideological boundaries between good and evil) for Bond's surveillance activities. Second, they provide opportunities for Bond to demonstrate his resources and wits and paradoxically perhaps his dependence at times on extraordinary gadgets such as the portable breathing apparatus in *Thunderball* (1965). The latter enabled Bond to escape the confines of a swimming pool filled with ravenous sharks. Third, they reaffirm the power of the individual to make a difference in a world made extreme by nuclear weapons and global tension. As the *Hollywood Reporter* reflected on the film, *You Only Live Twice* (1967):

> In a world traumatised by computerisation, explosive brush wars and the strained détente of the two major world powers, James Bond films offers an outsized affirmation of the power of the individual. Fall-out and rocket-rattling are trifling compared to the immediate contest for world survival between the current mad man at SPECTRE and the hyperglandular [sic] representative in Her Majesty's Secret Service.[65]

Finally, they almost always feature in the final climaxes of the Bond films. As Cawelti and Rosenberg contend, 'their sagas move towards a climactic scene which usually takes place in an extreme situation or locale, such as a wilderness or in the midst of a battle … Fleming invented a variety of locales with a similar function [to Buchan's novels].'[66] Thus, many viewers might be disappointed if the Bond films climaxed in seemingly more humdrum locations such as a suburban shopping mall.

Ironically, despite Bond's inevitable triumph, many of the stories end up highlighting the profound vulnerability of Britain and the West during the Cold War. Despite the technological and scientific resources of the United States and its allies such as Britain, Western civilisation remains vulnerable to Soviet plotting and criminal networks such as SPECTRE, which remain addicted to global domination. Evil geniuses such as Dr No, Goldfinger and Blofeld provide further proof that danger is never far removed from imperial regions such as the Caribbean and even the domestic interior of the United States. As with the post-11 September era, the apparently easy availability of nuclear weapons calls into question how highly industrialised societies such as the United States and Britain can be imperilled by the very technologies that seem to offer the illusion of complete security. There is nowhere that appears to be safe.

CONCLUSIONS

The James Bond films were remarkably successful in the 1960s in terms of audience figures and box office receipts. As the *Guardian* film reviewer noted in November 1964, 'the story [in this case the film *Goldfinger*] is fantasy, and fantasy is the stuff of cinema'.[67] Even if the trade press reviewers differed over their fantastical qualities, Ian Fleming's super-spy clearly reflected not only the prevailing Cold War culture but also the changing domestic and international profile of Britain. Fantasy was anchored to political detail and geographical background, which added an air of realism to Bond's missions and encounters. Regardless of whether the filming process took place on location or was 'faked' in Pinewood Studios (Goldfinger's Kentucky ranch was actually filmed in RAF Northolt in North London), places such as Turkey and the Caribbean played a significant role in conveying Cold War intrigue and providing a stage for the evolving Anglo-American 'special relationship' personified by Bond and Leiter. But as we have also seen, particular types of places (rather than geographical locations per se) perform an important narrative and visual role in the James Bond film series especially the use of inaccessible and or underground bases and or hideouts to generate a sense of dangerous climax.

More broadly, by exploring the role of places in films we might comprehend better some of the phenomena that make up a popular geopolitics. Three final thoughts can be offered at this stage. First, within films places provide opportunities not only to locate activities but also to shape narrative structures. In other words, places are rarely simply passive backdrops. They help to shape story lines and, in the case of political thrillers such as the James Bond film series, they often function to draw attention to prevailing dangers and threats. Locations such as the Middle East and the former Soviet Union in the aftermath of the Cold War have been particularly popular with Hollywood producers more generally. Second, the very nature of the film also provides opportunities to frame particular spaces. The fleeting images of London and the close ups of M's office in London can be considered in juxtaposition to the more elaborate 'field locations' and thus help to contribute to a sense of how danger is to be found within the exotic and or the foreign. The domestic functions as a metonym for stability and predictability. Third, there is a broader history and geography to be drawn with regard to the symbolic and material significance of tunnels and other confined spaces (and associated modes of travel) not least with reference to Cold War espionage.

Popular geopolitics, thus far, has not really considered in sufficient detail the nature of the filming process and the ways in which the very making of the film creates these different spatial settings. There is a great deal more to be done in showing how popular and political cultures interact with one another and associated geographical representations of global political space. With the ongoing intelligence debacle regarding Iraq's supposed nuclear weapons arsenal and international terror networks, there may be even more opportunities to consider how films involving the exploits of spies informs, contests and or reinforces politically hegemonic views of global political life.

ACKNOWLEDGEMENTS

I would like to thank Matthew Jones and Jeffrey Richards for their comments on earlier versions of this essay as well as the participants at the Britain and the Culture of the Cold War Conference 12–13 September 2003 and the participants at the 'Geopolitics' sessions of the SGIR Conference, 9–11 September 2004. My thanks to Tim Cresswell and Marcus Power for their support and aperçus. Amy Cooper of the University of Iowa helped greatly with accessing some of the Richard Maibaum papers. Bridget Robison provided invaluable research assistance. David Newman and the referees attached to *Geopolitics* offered many useful suggestions.

NOTES

1. Sir Nicholas Henderson interviewed by M. McBain, 24 September 1998, Churchill College Archives DOHP 32, p.17.

2. J. Cawelti and B. Rosenberg, *The Spy Story* (Chicago, IL: University of Chicago Press 1987) p.128.

3. For two recent if different examples on how International Relations scholars have explored how film might complicate our popular and theoretical understandings of inter-state relations, see R. Gregg, *International Politics on Film* (Boulder, CO: Lynne Rienner 1998); and C. Weber, *International Relations Theory* (London: Routledge 2002). Alternatively, on television programmes such as *Star Trek* see J. Weldes, 'Going Cultural: *Star Trek*, State Action and Popular Culture', *Millennium* 28 (1999) pp.117–34.

4. Weber (note 3) p.9.

5. In the sense that geographers have had a lengthy interest in the visual media such as M. Clark and W. Allen, 'Films: Cities via the Screen', *Geographical Magazine* 49 (1977) p.341. Earlier work included G. Cons, 'The Geographical Film in Education', *Geographical Magazine* 32 (1959) pp.456–66 and W. Harris, 'Using Motion Pictures to Illustrate Certain Aspects of Economic Geography', *Journal of Geography* 63 (1964) pp.72–3.

6. For an important intervention in the field of geography and music see J. Connell and C. Gibson, *Sound Tracks* (London: Routledge 2003).

7. Within human geography, there has been at least 20 years of scholarship concerned with the critical analysis of film and television. See, for example, S. Aitken and L. Zonn (eds), *Place, Power, Situation and Spectacle: The Geography of Film* (Trenton, NJ: Rowman and Littlefield 1993).

8. M. Heffernan, 'Balancing Visions: Comments on Gearoid O Tuathail's Critical Geopolitics', *Political Geography* 19 (2000) pp.351–4.

9. For example, J. Sharp, *Condensing the Cold War* (Minneapolis, MN: University of Minnesota Press 2001); and others such as E. Berg, 'Some Unintended Consequences of Geopolitical Reasoning in Post-Soviet Europe: Texts and Policy Streams, Maps and Cartoons', *Geopolitics* 8/1 (2003) pp.101–20.

10. Sharp (note 9).

11. See, for example, T. Cresswell and D. Dixon (eds), *Engaging Film* (Totowa, NJ: Rowman and Littlefield 2002).

12. See, for example, L. Giannetti, *Understanding Movies* (Harlow: Prentice Hall 2000) pp.66–70.

13. M. Denning, 'Licensed to Look', in C. Lindner (ed.), *The James Bond Phenomenon* (Manchester: Manchester University Press 2003) p.66.

14. F. Moretti, *Atlas of the European Novel 1800–1900* (London: Verso 1998).

15. L. Benton, 'Will the Real/Reel Los Angles Please Stand Up', *Urban Geography* 16 (1995) p.160.

16. Ian Fleming was married to Ann, the former wife of Lord Rothermere, the owner of the *Daily Mail*. Despite their tempestuous marriage, his Jamaican home *Goldeneye* was an important hub in the Fleming's international social network and welcomed regular visitors such as Noel Coward. They also met, in Jamaica, the former British spy and internationally acclaimed writer Graham Greene. Most famously, Anthony and Clarissa Eden sought refuge at the house in the disastrous aftermath of the 1956 Suez Crisis. As Ann Fleming recorded in her letter to Evelyn Waugh on 24 November 1956, 'The departure of the Edens for *Goldeneye* was wrapped in mystery ... Ian went to see Clarissa the night before they left, he found her practising with an underwater mask in her drawing room, no doubt part of the delusion that the Suez Canal was running through it'; see M Amory (ed.), *The Letters of Ann Fleming* (London: Collins Harvill 1985) p.189.

17. See, for example, A. Lycett, *Fleming* (London: Phoenix 1995).

18. On teaching James Bond and popular geopolitics, see K. Dodds, 'Reel Geopolitics', *Media Education Journal* 34 (2003) pp.3–7.

19. A point made by J. Black, *The Politics of James Bond* (New York: Praeger 2001), for example.

20. In the 1950s and 1960s, sex and spies made for a heady combination and President Kennedy was not the only public figure to be fascinated by such a combination. Audiences more generally wanted, according to Cawelti and Rosenberg (see note 2), to be entertained by the heroic adventures of a gentleman spy rather than the cynical and clandestine missions penned by Eric Ambler and Graham Greene. But no assumption is made in this essay that audiences all interpreted the films in a similar manner. Moreover, no attempt is made to recover the original intentions of the author, Ian Fleming. If nothing else, such a gesture would seriously underestimate the role of screenplay writers and others in 'converting' the novels to celluloid.

21. Despite his declining health, Fleming was the guest of the Kennedy brothers at a dinner in the Georgetown district of Washington. Over dinner, John Kennedy asked Fleming how the United States might tackle the problem posed by Fidel Castro's Cuba. The Kennedy administration was obsessed with the removal of the socialist president. According to Fleming, the use of 'ridicule' was deemed essential especially in a culture where only 'money, religion and sex' made an impression on Cubans (and perhaps the Kennedys as well). Fleming proposed that Kennedy should drop over the island thousands of fake dollar bills and leaflets declaring Castro to be sexually impotent. While his ideas were not followed up, Kennedy did order the CIA to investigate the possibility of producing either a poisoned cigar and or exploding shellfish that might be used to assassinate Castro. Allen Dulles, the then head of the CIA, asked without success to see Fleming before his departure for Jamaica. Dulles wanted to explore further his suggestions for the overthrow of Castro. In April 1961, the CIA's covert activities in Cuba were given a fatal blow as the proposed invasion by Cuban exiles at the Bay of Pigs turned into a fiasco. Kennedy was later to be assassinated by the Bond fan, Lee Harvey Oswald. Kennedy never got to watch his ninth favourite novel translated into film, because *From Russia with Love* was not released in the United States until April 1964.

22. For some important academic interventions relating to the political and cultural trajectory of James Bond see Lindner (note 13).

23. T. Shaw, 'The Politics of Cold War Culture', *Journal of Cold War Studies* 3 (2001) p.59.

24. E. Thomas, *Robert Kennedy* (New York: Simon and Schuster 2000) p.119.

25. President Clinton apparently liked to watch the film *High Noon* (1965) whilst occupying the White House. I owe thanks to Gertjan Dijink for this observation.

26. There is a substantial literature on Bond including major monograph length treatments such as T. Bennett and J. Wollacott, *Bond and Beyond* (London: Macmillan 1987); J. Chapman, *Licence to Thrill* (London, I.B. Tauris 1999); and Black (note 19). Others such as Umberto Eco have explored in great detail the narrative structures (Manichean divisions between Bond and the villain for example) underpinning the James Bond novels.

27. David Cannadine has noted that Britain was often described as a 'nation in decline' as evidence accrued of imperial dissolution in the tropical and temperate empire on the one hand and a much debated relaxation of public morality and probity on the other hand. As Bond retorted to his Japanese colleague, Tiger Tanaka in *You Only Live Twice*: 'England may have been bled pretty thin by a couple of World Wars, our Welfare State politics may have made us expect too much for free, and the liberation of our Colonies may have gone too fast, but we still climb Everest and beat plenty of the world in plenty of sports and win Nobel Prizes ... But there is nothing wrong with the British people – although there are only fifty million of them'. The impetus for such a stout defence of post-war Britain had been provided by an earlier aside by Tanaka on the implications of the 1956 Suez Crisis, which witnessed the political and financial humiliation of Britain at the hands of the Eisenhower administration. As a consequence, British prime ministers from Macmillan onwards sought a 'special relationship' with the United States in an attempt to restore a sense of national prestige and international purpose. It was, as Cannadine contends, an ideal moment to write about a character who would offer the reader a heavy dosage of 'great power nostalgia, imperial escapism and national reassurance'. see D. Cannadine, *In Churchill's Shadows* (London: Allen Press 2002) p. 279–311.

28. See J. Chapman and M. Hilton, 'From Sherlock Homes to James Bond: Masculinity and National Identity in British Popular Fiction', in S. Caunce et al. (eds), *Relocating Britishness* (Manchester: Manchester University Press 2004) pp.126–47.

29. C. Bold, 'Under the Skirts of Britannia: Re-reading Women in the James Bond Novels', in Lindner (note 13).

30. M. Denning, *Cover Stories: Narrative and Ideology in the British Spy Thriller* (London: Routledge 1987), especially the chapter 'Licensed to Look'.

31. C. Baron, 'Doctor No: Bonding Britishness to Racial Sovereignty', *Spectator: USC Journal of Film and Television Criticism* 14 (1994) pp.68–81.

32. T. Balio, *United Artists* (Madison, WI: University of Wisconsin Press 1987) p.253.

33. The film rights to *Casino Royale* had been sold to Gregory Ratoff in 1955 for only $6,000. CBS had earlier purchased the rights to televise *Casino Royale* and broadcast in October 1954 an hour-long programme on its 'Climax Mystery Theatre'. Ratoff was never able to find any financial backers and his widow later sold the film rights to Charles Feldman in 1960. Interestingly, the televised version of *Casino Royale* featured an American Bond and a British Leiter (renamed Clarence). Thanks to Jeffrey Richards for making this point to me.

34. See P. McGilligan, 'Richard Maibaum: A Pretence of Seriousness', in P. McGilligan (ed.), *Backstory: Interviews with Screenwriters of Hollywood's Golden Age* (Berkeley, CA: University of California Press 1986) pp.266–89.

35. A. Turner, *Goldfinger* (London: Bloomsbury1998) p.131. See also R. Sellers, 'Goldfinger', *Film Review* 21 (1997) 47–51.

36. Turner (note 35) pp.131–2.

37. McGilligan (note 34) p.286.

38. I. Fleming, *Dr No* (London: Penguin 2002) pp.212–14. The original was published in 1957.

39. Ibid. pp.236–7.

40. Papers of Richard Maibaum, MSc 149 Special Collection of the University of Iowa, Box 20 – A film treatment of *Dr No* dated 7th September 1961, p.27.

41. Papers of Richard Maibaum (note 40) p.36.

42. Papers of Richard Maibaum (note 40), A film treatment of Dr No dated 25th September 1961, p.46.

43. Papers of Richard Maibaum (note 40), Box 34 – untitled and undated paper.

44. Amory (note 16) p.297.

45. A. Marwick, *The Sixties* (Oxford: Oxford University Press 1998) pp.17–18.

46. 'Dr No', *Film Daily* 122/53 (19 March 1963) p.7.

47. 'Dr No', *Kine Weekly* 2866 (6 September 1962) p.15.

48. *Playboy* later became part of the widespread 'branding' of the James Bond films. For example, James Bond's wallet is shown to contain a membership card for the *Playboy* Club in *Diamonds are Forever* (1971).

49. Black (note 19).

50. In the novel *Dr No*, Fleming reflected on the nature of 'modern Jamaica' and complained that exclusive colonial clubs such as Queen's would not survive the post-colonial transformation. Ann Fleming, in one her letters to Eveyln Waugh dated February 1964, denounced the independence of Jamaica and complained that 'aggressive Negro custom officials enjoyed themselves hugely by a minute examination of underclothes'; Amory (note 16) p.333.

51. For a more scholarly reading of the US–UK intelligence relationship see R. Aldrich, *The Hidden Hand: Britain, America and Cold War Secret Intelligence* (London: Routledge 2001).

52. Prime Minister Anthony Eden famously took a first in Oriental Languages at Christ Church Oxford. Oriental Languages in this context referred to Arabic and Persian. I owe this observation to Matthew Jones.

53. I. Fleming, *You Only Live Twice*, p.35.

54. 'You Only Live Twice', *Variety*, 14 June 1967.

55. F. MacDonald, 'The Atlantic as Perceptual Field: Rockall and the Cold War', paper presented to the 2003 RGS-IBG Annual Conference London, 4–7 September 2003.

56. Quoted in K. Philby, *My Silent War* (London: Panther, 1973) p.142. Wisner was referring to Malta, which proved a useful base to launch espionage operations against the Albanian regime.

57. C. Hitchens, 'How does Bond Keep it Up?', *The Observer*, 17 March 2002.

58. This does complement the description offered of SPECTRE in the novel *Thunderball* (Harmondsworth: Penguin 2002) pp.62–4. The original was published in 1961.

59. Anglo-American suspicion of de Gaulle's France was well entrenched by the mid 1960s after the French decision to withdraw their Mediterranean fleet from NATO control, their advocacy of joint control of any US nuclear weapons stationed on French soil and France's decision to establish diplomatic relations with China in 1964.

60. For an extended discussion of Dr Fu-Manchu see L. Ling, *Postcolonial International Relations* (Basingstoke: Palgrave 2001) pp.81–97.

61. See K. Dodds, 'Licensed to Stereotype: James Bond, Popular Geopolitics and the Spectre of Balkanism', *Geopolitics* 8/2 (2003) pp.125–54.

62. Perhaps it would be more appropriate to say M's office in London, which for Bond is reassuringly unchanged for much of the Bond film series. As befitting M's nautical background, the office is filled with naval and imperial memorabilia. See P. Stock, 'Dial M for Metonym: Universal Exports, M's Office and Empire', in C. Landner (ed.), *The James Bond Phenomenon* (Manchester: Manchester University Press 2003) pp.215–31.

63. See the useful analysis by Baron (note 31).

64. See D. Stafford, *Spies Beneath Berlin* (London: John Murray 2002).

65. 'You Only Live Twice Still Another James Bond Hit', *Hollywood Reporter*, 15 June 1967, p.3.

66. Cawelti and Rosenberg (note 2) p.127.

67. 'Goldfinger', *The Guardian*, 18 September 1964.

Confronting the Geopolitical Aesthetic: Fredric Jameson, *The Perfumed Nightmare* and the Perilous Place of Third Cinema

DEBORAH DIXON
Institute of Geography and Earth Science, University of Wales-Aberystwyth, Aberystwyth, Wales, UK

LEO ZONN
Department of Geography and the Environment, University of Texas at Austin, Austin, Texas, USA

INTRODUCTION

In *The Geopolitical Aesthetic* (1992), Fredric Jameson effectively re-envisioned cinema as a complex 'space' that engages the desires and concerns of the individual subject and provides them a window through which they can glimpse the social totality within which they, and the film itself, are embedded. In this sense of the term, the geopolitical aesthetic is at once a product of what Jameson views as the prevailing social 'fact' of our time, late capitalism, and a vehicle through which the individual can come to reflect on, consciously or unconsciously, this very same reality. Combining a Hegelian Marxist understanding of the role and function of cultural texts within the economy, with a utopianism regarding the potentially progressive engagement individuals can achieve with such texts, Jameson's writing on cinema offers an avowedly spatially sensitive framework for those working on film, but also an opportunity to think about how concepts of space and scale can be actively deployed in more general accounts of globalisation, within and beyond the discipline of Geography.[1]

In the following we provide a brief summary of the transformative impact of Jameson's (1992) writing on cinema, focusing in particular on how his notion of the geopolitical aesthetic effectively works to undercut the category of Third Cinema (long extolled as a site of resistance) as a distinctive and legitimate object of analysis. We then proceed to critically examine the geographies that animate Jameson's analysis, focusing on his reading of Kidlat Tahimik's *The Perfumed Nightmare* (1976). Extolled by some as a classic of Third Cinema because of its critical stance in regard to American imperialism in the Philippines,[2] Jameson uses the film to make the argument that Third Cinema is very much a redundant force within late capitalism; instead, he advises, we must look to explicitly postmodern forms for resistance to reification and commodification.

While we find much that is productive in Jameson's writing, we would like to point to the conceptual limitations at work in Jameson's reading of *The Perfumed Nightmare*, which is key to his concept of the utopic character of film more generally and, moreover, to his argument on the embeddedness of Third World representations within a global, capitalist system. This is because although Jameson (1972) acknowledges the underlying constructed and relational character of ontological categories such as film (despite their reification under capitalism), his theory of historical materialism demands that they also be understood as formed with regard to a socio-economic totality. And, because the recognition of a totality requires a master narrative within which all can be understood and framed within a logic of equivalence, Jameson must by default conceive of epistemology as fundamentally divided between a true and a false consciousness.

Taking our own cue from recent developments in anti-essentialist thought, we conceive of such cultural forms as the temporarily fixed embodiment of broader-scale discourses that continually construct and deconstruct the world as we know it, including our understandings of the 'real' as well as 'representation', the 'economic' as well as the 'cultural', the 'social' as well as the 'subject' and the 'local' as well as the 'global'. Put briefly, within anti-essentialism attention is drawn to the active and continued construction of difference through epistemological categories such as the certain/real/material/true and the uncertain/ghostly/imaginary/false:[3] Derrida's 'constitutive outside' refers to the fact that this naming process relies on a series of negative distinctions that define a 'thing' (whether it be an idea or a designated object) according to what it is not. Hence, the idea/object under scrutiny is by its very existence related to a host of other ideas/objects. Moreover, when the naming process is effective to the point that the meaning of an idea of object becomes commonly accepted, we have what Barthes calls 'myth', that is a construction that is taken for granted, or 'necessary'.

Such scepticism extends unto theories of a social totality, or world system. Indeed, as has been argued by the proponents of 'critical' geopolitics, it is in part through the promulgation of a naive form of realism that 'nation-states' have been afforded such an authoritative status as the central agents of global-scale, power relations, such that one can talk of 'statecraft' and 'foreign policy'. In similar vein, it is partially through Marxism, it can be argued, that the capitalist economy continues to be rendered legible through the designation of periods, sectors, classes, occupations, crises and solutions. As a counterpoint, partial genealogical and deconstructive analyses can explore the constructed character of this pillar of domination.[4] Given our own anti-essentialist perspective, it is not surprising that we find Jameson's notion of cognitive mapping doubly problematic, first, on the grounds that it presupposes an universal search for the 'real', as constituted by late capitalism, and, second, because it posits that reality as existing at the 'global' level.

From this specific vantage point, therefore, we hope to make clear how Jameson's understanding of the geopolitical aesthetic, and the specific practice of cognitive mapping, leads him to read *The Perfumed Nightmare* as a mode of resistance over and against late capitalism, and to locate and frame particular spaces within the film as meaningful and significant in regard to this project. From our perspective, not only does this reading of the film render other theorisations of film as limited, and hence unproductive, concerns, it also ensures that Jameson's accounting of those spaces he does recognize — namely, the economic, the political and the cultural — remains anemic. In our own assessment of *The Perfumed Nightmare*, we provide a partial response to this, noting how these spaces are constituted from the temporary 'fixing' of a series of people and place-based identities, such as those constituted under the rubric of 'gender'. This requires much more than a simple acknowledgement of particular narrative themes that invoke gendered characters, such as son, daughter, wife, husband, mother and so on, but an exploration of the ways in which matters of economic, political and cultural import are imbricated throughout with notions of what it is to be feminine and masculine.

THIRD CINEMA OR GEOPOLITICAL AESTHETIC?

In this section we introduce the concept of Third Cinema, drawing out how this genre has been designated by a number of film theorists as 'progressive' according to the conditions of its production, as well as its style and form. We then go on to introduce the work of Fredric Jameson, noting how his (1992) work on cinema must be understood as part of a broader-scale project to describe and appraise what is termed the 'postmodern condition' under late capitalism. Jameson's work directly challenges the idea of Third Cinema film as an object of study because he regards film in general as a potentially progressive medium through which late capitalism can be problematised. All such cultural texts allow for what Jameson has termed cognitive mapping, whereby the individual subject is able to experience a sense of the social totality within which they are placed.

The term 'Third Cinema' was coined by Argentinean film-makers Fernando Solanos and Octavia Getino in 1969. Invoking the Cuban Revolution of 1959 as well as a post-1968 European and American avant-garde film school, Solanas and Getino extolled Third Cinema as an artistic and social movement hostile to the dominance of an industrialised, ideologically corrupt Hollywood, as well as a politically inept 'author-cinema' movement located in the First and Third World.[5] By contrast, a 'guerrilla' cinema would work to de-colonise cultures: 'The camera is an inexhaustible expropriator of image-weapons; the projector, a gun that can shoot 24 frames per second'.[6]

Solanas and Getino also emphasise the many forms that Third Cinema can take, such that there are no prescriptive methods advocated as to how such films are produced (1973). This is because, as Solanas makes clear, Third Cinema *is* defined according to who produces it, in that it is the medium of the 'people', comprised of industrial workers, small and landless peasants, the unemployed, the lumpenproletariat and students.[7] (From our perspective, such a stance necessarily raises the question not only of how such neat demarcations are to be made and sustained, but also how Third Cinema films are understood to retain their ideological credentials when viewed by those who do not share similar presumptions?)

While the writings of Solanas and Getino have formed the basis of a wide range of essays and commentaries on Third Cinema published in Latin America and North Africa,[8] the concept is most closely linked with the writings of Teshome Gabriel, an Ethiopian filmmaker, on the aesthetics of liberation. For Gabriel, it is immaterial as to who produces Third Cinema, for this is a genre characterised by its perceived goals:

> The principal characteristic of Third Cinema is really not so much where it is made, or even who makes it, but, rather, the ideology it espouses and the consciousness it displays. The Third Cinema is that cinema of the Third World which stands opposed to imperialism and class oppression in all their ramifications and manifestations.[9]

For Gabriel, film has the capacity to reflect on the conditions of its own production, thereby serving as a means of raising class consciousness. Via its own particular codes and conventions, Third Cinema can communicate an ideological position that undercuts the normative Western stance of First World films. For example, Gabriel points to how high/low camera angles can emphasise the socio-spatial disparities of subjects; the use of wide-angle shots can emphasisze the communitarian context of subjects; straight-to-camera dialogues can bridge the subject-viewer divide; the juxtaposition of times and space via dynamic editing can delineate socio-spatial contrasts between groups; and the dissonant overlay of non-synchronised English over indigenous languages can critically portray the colonial imposition of cultures.

In emphasising the fact that film, no matter where or by whom it is produced, can provide a critical commentary on colonialism and neocolonialism, Gabriel effectively 'globalises' the concept of Third Cinema. In doing so, however, he presumes a global-scale economic system that not only embeds the coloniser and the colonised, but which also provides the primary means of identification for these subjects. This is because although gender, sex, race, nationality and culture are all recognised as key thematics of a Third Cinema film, each is understood to be articulated through class antagonism. For film scholars and practitioners in the 1980s, however, the point of contention within Gabriel's formulation lay not in his reading of the 'reach' of the global economy, but rather in his prioritisation of class above all else. In particular, how did Third Cinema intersect with Third *World* cinema, and particularly those films associated with the articulation of a national identity?

In 1986, the Edinburgh Film Festival hosted a three-day conference on precisely this question, bringing together Anglo and American film scholars with those referred to in the conference booklet as 'cultural activists' from Asia, Africa and Latin America. The conference was conceived of as an opportunity to explore whether or not an international tradition of filmmaking did indeed exist that 'exceeds the limits of both the national-industrial cinemas and those of Euro-American as well as English cultural theories'.[10] It is also important to note that for the conference organisers this was an opportunity to debate and reject what was seen as a 'reactionary' postmodern celebration of popular culture. The presentations were subsequently published as an edited collection under the title of *Questions of Third Cinema*.[11]

Subsequent writers have expanded on the emancipatory character of Third Cinema.[12] The utility, and indeed the legitimacy of this object of analysis has, however, come under sustained critique from one of social theory's preeminent writers, Fredric Jameson. Jameson began his publishing career as a Marxist literary critic: *The Origins of a Style* (1961) focuses on what was to become an enduring theme, the issue of narrative, while *Marxism and Form* (1971) concentrates on key figures of Hegelian Marxism, including Adorno, Benjamin, Bloch, Lukacs, Marcuse and Sartre. Jameson's deployment of a critical Marxist tradition has been complicated, however, by an engagement with continental structuralist and poststructuralist linguistics, notably in *The Prison-House of Language* (1972).[13] Though the specific cultural texts Jameson chose to work with early on were predominantly literary, he has also built up an extensive publication record on 'popular' cinema. These essays are thoroughly indebted to the Frankfurt School,[14] and include a series of commentaries on the politics of select films, including *Zardoz*,[15] *Dog Day Afternoon*[16]and *The Shining*,[17] as well as the deployment of magical realism in cinema.[18] Much of this work was republished in *Signatures of the Visible* in 1990.[19] Jameson's active research of that year, however, centred on the question of how cinema as a 'postmodern' phenomenon could be considered 'progressive'. In May 1990, Jameson presented a series of lectures to the British film Institute on this issue. As a means of illustrating his argument, Jameson used, amongst other material, Kidlat Tahimik's *Perfumed Nightmare*.[20]

The arguments of *The Geopolitical Aesthetic* must be understood as embedded in a broader-scale formulation on the relationship between the individual and the social, as articulated by Jameson at a particular stage in his academic career. Put briefly, Jameson takes as given Ernest Mandel's thesis that we are experiencing the late or 'third stage' of capitalism characterised by multinational capital. Eschewing a simplistic notion of base and superstructure, wherein the Economic is rendered as a hidden Essence, Jameson deploys Althusser's notion of a totality to argue that each stage has its own cultural styles — realism, modernism and postmodernism – which, although they are dialectically related to the mode of production of their time, have their own distinct logic. Moreover, each cultural style has its own spatial sensibility, as realism works to the grid, modernism thrives on contradiction and postmodernism dissolves into hyperspace.[22] Importantly for Jameson, it is precisely because these cultural styles are socially produced that they allow people the means of reflecting on the character and consequences of late capitalism. Though particular cultural texts, such as film, may not provide such critical comments directly, and may even promote a bourgeois ideology, they articulate a link between the individual and the social, and so allow for the everyday person to evaluate the prevailing political economic landscape and recognise their position within it, even if unconsciously so.[23] It is this fumbling towards realisation through the realm of the cultural that Jameson terms 'cognitive mapping'.[24]

Rather than bemoan what he sees a postmodern rejection of metanaratives, therefore, Jameson perceives in such 'loss' and 'confusion' a utopian desire. That is, the feeling of loss serves to highlight the pleasurable yet imaginary unity of a social collective: this imagined unity, in turn, 'is utopian not as a thing in itself, but rather to the extent that such collectivities are themselves ciphers for the final concretion of collective life, that is the achieved utopia of a classless society'.[25] Importantly, the form of these utopian imaginings – that is, their postmodern character – is constituted from a deep-seated aversion to the current state of affairs, as organised under late capitalism. Echoing the work of Benjamin on film,[26] Jameson concludes that each postmodern cultural object contains within its very production, 'our deepest fantasies about the nature of social life, both as we live it now, and as we feel it in our bones it ought to be lived'.[27]

The analytic framework Jameson has developed in these texts – wherein Marxism provides the overarching theoretical framework through which issues of class, struggle and representation can be explored – has profound implications for Third Cinema as an analytic category and distinct object of study. Because all films have a utopian moment, Jameson argues, then all must be considered potential modes of resistance through the practices of production and consumption. Hence, while Jameson also 'globalises' cinema, he does so by undercutting Gabriel's notion that films can be labeled as 'progressive' according to the *intent* of their producers. Thus, for Jameson, there is no Third Cinema as such, just as there is no First or Second Cinema. Instead, he refers to Third *World* Cinema, and the productive potential thereof, a phrasing that challenges us to reconsider the 'spaces' within which film is produced.

In view of the above it is interesting to note why Jameson is so taken with the work of Filipino director, producer and actor Kidlat Tahimik. He appreciates Tahimik's *The Perfumed Nightmare* not because it is somehow representative of Third Cinema as a genre, but because he reads Tahimik's project as a particular instance of cognitive mapping. For Jameson, Tahimik's film is embedded within but also transcends his local and regional cultural context as the director strives to confront and critically meditate on First World economics, technologies and aesthetics.[28] Thus, the film affords Jameson an opportunity to discuss how a film produced within the 'Third World' does not simply 'represent' that context in particular ways, but is constituted in large part through its deployment of symbols, allegories and techniques that invoke a sense of the global. What makes Tahimik's film so distinctive for Jameson is the fact that while most modernist, Third World films tend to focus on the social disjunctures created by colonialism and neocolonialism, or work unconsciously to invoke a 'geopolitical aesthetic' through their portrayal of 'archaic' categories such as nationality and myth, *The Perfumed Nightmare* is very much a self-conscious exercise that seeks to make a connection between the localised experience of the *individual* and the globalised totality that is late capitalism.

CONFRONTING THE NOTION OF GEOPOLITICAL AESTHETICS THROUGH *THE PERFUMED NIGHTMARE*

Before launching into our in-depth assessment of Jameson's reading of *The Perfumed Nightmare*, as well as our critique of the same, it is as well to briefly introduce the film and its maker. Kidlat Tahimik (aka Eric de Guia) was born in the Philippines in 1942, was educated at the University of the Philippines, and then moved to the United States for his MBA at the Wharton School of Business. Following a brief stay in Europe, Tahimik returned home, whereupon he soon came to the realisation that he had grown up 'sleeping in the cocoon of his American dreams' and later tore up his American diploma, because 'it's a part of my past. It's not relevant to me.[29] His first feature work, *Mababangong Bangungot (The Perfumed Nightmare)*, is a semi-autobiographical film that he wrote, produced and directed, and in which he played the protagonist, named Kidlat Tahimik (which translates as 'quiet lightning'). The film was independently financed, primarily by Tahimik himself working as a jeepney driver. and was produced for US$10,000 at a time when the average cost of a Filipino film was US$100,000. Originally made in Super 8mm, the film generated considerable critical acclaim, winning awards at several international film festivals, including the 1977 Berlin Film Festival International Critics Award. The film has since become something of an art-house classic. And yet, as Sison points out, the film was not a commercial success in the Philippines itself.[30]

The actual narrative of *The Perfumed Nightmare* is fairly straightforward. Kidlat makes a living as a jeepney driver, and in his spare time is president of the Werner von Braun Club of Balian. In the first part of the film, during which we follow Kidlat's journeys around, to and from the village with an assortment of passengers, it is made clear that Kidlat idolises other aspects of Western, particularly American, culture, including the Voice of America, the Statue of Liberty and the Miss Universe Beauty Pageant. Kidlat takes a job as chauffeur with a visiting American and moves with him to Paris (with the jeepney in tow). Much of his time is spent refilling his boss' gumball machines, which are scattered around Paris. Kidlat also becomes friends with a local market stallholder, whose livelihood is threatened by the construction of a gigantic *supermarché*. A trip to Munich – the birthplace of Werner von Braun – furthers his growing disillusionment with Western technology, as Kidlat witnesses the last, handcrafted *Zweibelturm* being hoisted into position. This, together with the news of his mother's hut being displaced by the construction of a highway for tourists, provokes a wholesale rejection of an inhuman, Western zeitgeist and, in a magic realist assemblage of images, Kidlat blows away various symbols of Western, technological superiority. Kidlat's actions here are a homage to his father's exploits during the Philippine–American War, wherein, before being shot dead on the San Juan bridge, he blows away 15 US troops with the force of a typhoon. The film ends back in the Philippines with a shot of his mother slowly closing the window of her bamboo hut.

Jameson's *Perfumed Nightmare*

Before undertaking his own assessment of *The Perfumed Nightmare* Jameson makes clear his disappointment with avowedly 'progressive' Third World films, because, he suggests, late capitalism has ensured that such efforts have either been co-opted by an 'enlarged and ecumenical Hollywood cinema', or rely on the 'sterile passions of nationalism and religious fundamentalism'.[31] In looking to the Third World as a 'social space' from which alternatives to corporate, capitalist daily life are to be located, however, one might consider examples of 'postmodern' cinema. In such forms, he argues, one can perceive a productive politics. Specifically, one can point to the presence of cognitive mapping, as such films invoke, largely unconsciously, a sense of confusion and disorientation under late capitalism itself. It is the *conscious* instance of cognitive mapping that Jameson finds so appealing within *The Perfumed Nightmare*: as Jameson notes,

> it is obviously encouraging to find the concept of mapping validated by conscious artistic production, and to come upon this or that new work, which, like a straw in the wind, independently seems to have conceived of the vocation of art itself as that of inventing new geotopical cartographies.[32]

Importantly, Jameson is convinced of the consciousness of this practice because, he concludes, the film not only effectively apes First World aesthetics (that is, the style and technique of French film-maker Jean Luc Godard as well as the Russian Sergei Eisenstein) through its explicitly 'naïf' production, but also undercuts First World theories on the relationship between culture (in the form of US cultural hegemony) and political power (in the form of US gunboat policy). Eschewing a naive First World belief in the ontologically distinct character of each, Tahimik, Jameson argues, refuses to settle on either as a set of causal processes. Instead, he argues, Kidlat Tahimik, as director and producer, uses both in allegorical fashion to reference how the film's central figure, Kidlat Tahimik as actor, reaches out from his 'place' within the Third World to the social totality of late capitalism via his interaction with these elements: 'the relevance of Tahimik's production for the contemporary (or post-contemporary) situation lies precisely in the way in which he eschews the political for the economic, and the thematics of power for that of reification'.[33] In short, Jameson is claiming that Kidlat Tahimik uses his localised cultural and political situation to make indirect, critical reference to the real, globalised world of the economic that effectively contains both First and Third Worlds.

This is necessarily a situated, subject-driven exercise in 'bridging' worlds, because 'what the First World thinks and dreams about the Third can have nothing whatsoever in common, formally or epistemologically, with what the Third World has to know every day about the First ... In the same way, the village, as it extends outwards to include Manila, and then Paris and Europe itself, is a very different kind of space from that – exactly coterminous and identically superimposable on the same map – which stretches out from Paris and Europe to envelop the Philippines, Manila and ultimately the village itself'.[34] What this means, of course, is that just as Kidlat Tahimik practices his own cognitive mapping within the film, we, as viewers, undertake the very same exercise through our spectatorship. Hence, the film works to a two-fold geopolitical aesthetic through its production and consumption.

One might assume from the sequential placement of the film's protagonist in a series of increasingly 'First World' spaces – from Balian to Manila, Paris and Munich – that the film is actually 'modernist' in its representation of space, wherein each locale is rendered ontologically distinct. But, Jameson argues, this is a thoroughly postmodern representation in that each space is inextricably linked with the others through the symbol of the bridge. The bridge that links Kidlat's village to the rest of the world appears at the beginning of the film, as Kidlat Tahimik walks across it three times in succession, towing a progressively larger jeepney. Throughout the film reference is made to efforts at 'bridging' – rural and urban, Earth and moon, nature and culture – while examples of various bridges are shot *in situ* or as pictures within other texts. It is worth quoting Jameson at length here on the significance of this 'archaic' symbol to the representation of space in the film, as:

> the picture postcard of the bridge leads us further on into sheer space: the space of the village, and then the space of the bridge of transport between the village and Manila – figured by the jeepney that conveys passengers back and forth. At length, in a larger opening, this is not merely the bridge between the earth and the moon (along with the Werner von Braun fan clubs that celebrate it), but that more tangible bridge which the protagonist will at length cross leading from Asia to Europe, from Third World to First and back, from Manila to Paris (and from Paris to the Rhine), and from a Philippino present to a traditional Parisienne past itself in the process of being obliterated by its own Common Market future. All of these spaces are then inconstant decomposition and modernization, including each other heterogeneously, in such a way that narrative progression becomes unthinkable, except as a bus ride, and we learn to substitute for it the discontinuous series of spatial exhibits that might be offered by a collection of snapshots.[35]

The bridge, then, does not work to take us from A to B in some logical, sequential progression either over space or through time, but rather serves to dissolve the self-contained character of each locale.[36] This experience of a 'hyperspace' is, of course, a key hallmark of the 'postmodern condition'.[37] Thus, for Jameson, the bridge itself operates as an effective postmodern symbol of the cultural impacts of late capitalism.[38]

There are, however, two spaces that hold a special significance for Jameson in that they both explicitly invoke the spectre of late capitalism, a mode of production that works to commodify all aspects of human subjectivity and experience. The first is the body of the protagonist himself, whom Jameson describes as 'first and foremost a clown'.[39] The figure of the clown, Jameson argues, allows us to reflect on our own discomfort with the modern, urban environment, as:

> the setting in motion of that world demand a certain resistance; its tactile exteriors lend themselves to exploration and articulation by way of the elasticity of the clown's body ... the ungainliness of the protagonist elicits, like two surfaces slowly beginning to lean towards each other, the inhuman unloveliness of the glass walls and décor.[40]

For Jameson, the film's reliance on occasional farce – as when Kidlat attempts to negotiate Paris' Charles de Gaulle airport – make sense as a form of cognitive mapping. This 'libidinal investment'[41] in the body of the clown is a key meditation on late capitalism, he argues, just as much as it is an aping of the comic aesthetics of Jacques Tati and Charlie Chaplin.

It is this same body that appears in an early scene, set within the context of growing up in Balian, that depicts ritual, male circumcision. For Jameson, this graphic, 'scandalous' episode is intended for voyeuristic First World audiences,[42] in that it presents them with an ironic *ressentiment* of Third World otherness.[43] While seemingly 'anomalous', then, to the humorous 'travelogue' that Kidlat has constructed, this scene, according to Jameson, is rendered legible when one appreciates how Kidlat has deliberately plays to the concerns and attitudes of the First World. Thus, the scene does not lay claim to a representation of the Real – that is, the revivification of native custom – but is postmodern in its re-circulation of images and which is, therefore, indicative of capitalism's commodification of subject identities.

The second space that Jameson finds particularly interesting is the jeepney factory. For Jameson, Kidlat Tahimik's inclusion of this space is crucial. It proves that *The Perfumed Nightmare* is not a 'classic' Third Cinema film, for there is no simplistic binary here between First World/Modern/Industrial/ Inhuman and Third World/Traditional/Agricultural/Communal. Instead, Jameson argues, Tahimik imbues the jeepney production line with a hybrid identity: 'the jeepney factory is a space of human labor which does not know the structural oppression of the assembly line or Taylorization, which is permanently provisional, thereby liberating is subjects from the tyrannies of form and of the pre-programmed'.[44] Because there is no separation of aesthetics from production, the workers are producing bricolage as opposed to commodities. Because the hand-crafting of the jeepneys involves the continuous recycling of parts, there is no destruction or waste. And, because the jeepney partakes of Filipino, American and Spanish legacies,[45] there is no 'authentic' culture to be commodified and sold. For Jameson, this is an instance wherein a utopian workplace is actually realised, at least on film. Moreover, it forms an allegory for the film itself, which is, 'just another jeepney, an omnibus and omnipurpose object that ferries its way back and forth between First and Third Worlds with dignified hilarity'.[46]

It is precisely this reading of this space that has most concerned some Filipino academics writing on Jameson.[47] For these, *The Perfumed Nightmare* falls squarely within the contexts of Filipino New and Alternative Cinema. The politically influential 'New Cinema' emerged in the 1970s and was characterised by an increased quality of production, a greater emphasis on social themes, and increased budgets. New Cinema reflected a developing social consciousness, dramatically changing social and political conditions, and increasingly conservative strictures of film censorship from within the Marcos government (1965–86), particularly in terms of political statements in film. Perhaps encouraged by the social and critical successes of New Cinema and yet discouraged by their larger budgets, a smaller and far less commercial Alternative Cinema, which included the short film, as well as documentaries and feature films, began to emerge.[48]

For Lim[49] and Tolentino[50] especially, Jameson's reading of *The Perfumed Nightmare* betrays a lack of knowledge (or 'native informant position') concerning this material history and the realities of the jeepney workplaces, which they describe as thoroughly alienating and industrialized. Sympathetic to these concerns, San Juan notes that Jameson's reading is a selective one that reveals his obsession with commodification.[51] For San Juan, the film's critique of US colonialism cannot easily be dismissed, and indeed can be more usefully read as a modernist account of resistance in the face of such aggression: 'what needs underscoring is the running commentary that (Kidlat's) father and millions of Filipinos refused to be bought for $12 million dollars – the price the US paid to Spain for ceding the Philippines at the Treaty of Paris'.[52] Interestingly, San Juan favours the work of David Harvey[53] over Jameson in regard to his explanation of the workings of global economy, because Harvey's 'socio-temporal dialectic' appropriately recognises that successful resistance can only ensue from place-based, class antagonisms. This does not suppose a categorisation of *The Perfumed Nightmare* as Third Cinema, however, because its realistic style of distanciation and dislocation is intermittent, and so a progressive message concerning national liberation is muted. In the context of Filipino New Cinema, then, *The Perfumed Nightmare* is something of a failure.

And yet, the title of most recent compendium of writings on the Filipino film industry, Rolando Tolentino's *Geopolitics of the Visible*,[54] explicitly echoes Jameson's two earlier books, *The Geopolitical Aesthetic* and *Signatures of the Visible*. What this signals to us is that although Jameson's utopian vision is derided in part as a flawed, Western Marxist framing of Tahimik's work, and film more generally, his basic argument concerning the character and scope of late capitalism, as well as its realization within the cultural realm, is accepted as foundational. For Tolentino, the term 'geopolitics' refers to a 'transnational cultural politics that effects the implementation of globalizing forces in the local national landscape, and demonstrates how the local might become a trope for situating past and ongoing globalization drives'.[55] In other words, the geopolitical aesthetic revolves around the systemic codification of the Real, whereby capitalism renders itself invisible and thereby unassailable.

Reliving *The Perfumed Nightmare*

As envisioned above, Jameson's geopolitical aesthetic, while capturing some notion of the inter-relationality of the economic, the political and the cultural, also works to differentiate these terms along the lines of scale, or 'reach'. This is because within Jameson's understanding of ontology the economic is held to exist at a broad, pervasive, 'deep' level within society, while political and cultural phenomena are held to lie at the surface, readily available to the individual for observation and comment. Though we cannot behold the precise workings of the economic, we can gain some notion of its effects by looking to the other two. Indeed, Jameson's choice of the title 'Beyond Landscape' for his introduction to *The Geopolitical Aesthetic* makes clear his understanding of a deep ontology that, though hidden by surface phenomena, can be indirectly accessed through critical reflection. Indeed, it is precisely because Jameson himself knows the reality of this ontological depth that he can recognize its antithesis, namely a postmodern 'surficiality.' As Jameson explains this trend, postmodern theory rejects the 'truth' of several depth models, including essence and appearance (framed as ideology or false consciousness), Freud's manifest and latent (framed as repression), authenticity and inauthenticity (framed as alienation) and signifier and signified (framed as fixed meaning).[56] In consequence, Jameson argues, postmodern *theory* (as opposed to postmodern cultural artifacts) posits our current condition as 'sheer heterogeneity, random difference, a coexistence of a host of distinct forces whose effectivity is undecidable';[57] for Jameson, this rejection of depth not only ensures a misguided celebration of superficial phenomena, it also commits postmodernist theory to a reactionary social politics.

Of course, the form and function of the 'problematic' afforded by this understanding of ontology is dictated by the terms of Jameson's broader argument, wherein postmodernist theory is itself framed as a cultural condition under late capitalism. And yet, when this 'postmodern sensibility' is understood as a form of anti-essentialist *analysis* as opposed to a mere *symptom*, it is possible to question Jameson's deployment of scale. Recall that for Jameson, the economic is part and parcel of a social totality: it is because it is such a spatially extensive – i.e., a global – series of social relations, knowledges and practices that late capitalism plays a key role in shaping all aspects of that society. Recognition of this global-scale entity is akin to apprehending the 'reality' of one's own place in the world.

In counterpoint, the argument has long been made that a Marxist understanding of the 'global' reach of capitalism makes an unwarranted distinction between the 'necessary' social fact of the economic and the 'contingent', 'localised' play of the political and the cultural.[58] According to Hutcheon, for example, this is not a neutral statement of fact, but is a consequence of the often unconscious ethnocentrism and phallocentrism (not to mention heterocentrism)' manifest in the particular, discursive articulation of the economic found more often than not within Marxist texts.[59] More pointedly, Gibson-Graham notes that Jameson's work is illustrative of a tendency within Marxism to render capitalism in phallocentric terms, as it 'saturates' and 'penetrates' all possible spaces and voids. His work presupposes, it is suggested, a world "rightfully-owned" by capitalism.[60]

As we noted in the introduction, it is not surprising that we too find Jameson's notion of cognitive mapping problematic, in that it presupposes an universal search for the 'real' and because it posits that reality as existing at the 'global' level. This does not mean, however, that we find Jameson's key argument, that Kidlat Tahimik consciously practices a form of cognitive mapping, to be incorrect. Tahimik may well have deliberately 'displaced' a critique of late capitalism onto a humorous meditation on American cultural and political imperialism. What we do suggest, though, is that Jameson's reading of *The Perfumed Nightmare* as cognitive mapping effectively dismisses as insignificant the geographies of gender, as represented within the film, which, we imagine, must have fallen somewhere within Jameson's framing of a local, cultural realm. In doing so, we pick up on Gibson-Graham's argument that it matters as to how categories such as the economic are articulated within particular contexts and that gender, alongside other identity markers such as class, race, nationality, sexuality and so on, plays a crucial role in this framing.

In the following, therefore, we would like to 'recover' the geographies of gender within *The Perfumed Nightmare* somewhat, not because we see it as having a causal efficacy, nor because we wish to argue that Kidlat Tahimik assigns it causality. Rather, we wish to suggest that Kidlat Tahimik's *articulation* of the economic, the political and the cultural is thoroughly gendered. In assessing the means by which Kidlat Tahimik constructs each, however, we must of necessity undertake our own highly selective reading of *The Perfumed Nightmare*, wherein particular moments are highlighted as significant and meaningful. We offer the following, therefore, with the proviso that these comments are not a *definitive decoding* of the text, but are a reflection on moments in the film that we can use to provide an alternative read to that provided by Jameson.

There are several representations of people and objects in *The Perfumed Nightmare* that we have chosen to explore in terms of the gendering of the economic, political and cultural geographies of the film: Kidlat's mother and sister, and the mythical figures of the butterfly and the white caribou. A few scenes associated with each are highlighted in tabular form as part of a broader discussion.

We begin with the figure of Kidlat's mother, who remains unnamed through the film and whose images appear at strategic moments, including the closing of the film (see Table 1).

The mother is first seen as she sweeps a dirt pathway near her bamboo hut and, in fact, she is never found beyond a short reach of her own home. This placement of the mother is in distinct contrast to the movements of Kidlat, who travels often in his jeepney – to and from the village to outlying areas, as well as to the city (Manila), and eventually to Paris – and to Kidlat's father, grandfather, and friend Kaya, who are the other male figures of the film.

These travels of the men involve work, and so there appears to be the construction of a simple economic binary between a feminine, 'traditional' world of 'housekeeping' and the more public, masculine world of 'work'.

TABLE 1 Kidlat's Mother

Event	Context/setting	Dialogue
Framing mother, Madonna and beauty queen	Kidlat kisses the photograph of a blond beauty queen on a poster, who he calls 'love of my life', then the camera shifts (and holds for a moment) to his mother sweeping outside, then shifts back to Kidlat cutting out the beauty queen photograph, back to the mother again, then centres on Kidlat as he places the photo in a picture frame alongside a Madonna; the camera then shifts back to the mother sweeping.	Begins with voiceover from Voice of America, referring to the Bicentennial and then the space programme. As Kidlat prepares the photograph for the frame, he is providing the dialogue from his letter to Voice of America,which asks: 'What were the first words your great American astronaut said when they landed on the moon', whereupon the camera shifts to his mother sweeping.
Kidlat's mother tells the tale of his father	Kidlat and his mother are walking, as she talks. There is then a pantomime and clown-like routine of a smiling man in a sled, pulled by toy-like wooden animals, as if he were a 'taxi' driver. The man seems to be singing, but eventually picks up a rifle and becomes more serious. The camera goes back and forth from the mother telling the tale to Kidlat to the routine of the father in the 'taxi', and eventually shows the mother going into a hut and returning with a small, carved wooden horse.	Kidlat's mother to Kidlat: 'Kidlat, your father was a happy taxi driver. He always sang as he worked. One day a smiling stranger gave him a rifle. "The bridge to your freedom, your vehicle to freedom. We will help you with your revolution against the Spanish Tyrants", said the smiling Americano. Your father stopped singing. Before in the revolution and the days of the Spanish were numbered; when the Spaniards surrendered, your father sang again. As he sang the sweet song of victory, the Americans were buying us … Your singing father tried to enter liberated Manila at the San Juan Bridge. He was stopped by an American sentry. It was his last song. From the butt of your father's rifle, I carved this horse. Killed for trespassing on US property was the military report attached to his corpse. For twelve million dollars, we became US military property'.
Kidlat's mother ends the story.	Kidlat's mother is seen resting on a mat, and then she gets up. The camera then centres on the village, with one small butterfly seen flying slowly, and then the camera returns to the mother, viewed from outside the hut, as she slowly approaches the screen window looking out toward the camera. The camera pauses, and then the mother slowly lowers the bamboo shade, hiding her from sight. The camera shifts to a full view of the hut, pauses for a moment, and then all is black, ending the film.	A traditional and lamenting song is played quietly in the background.

Nonetheless, several subsequent portrayals undermine such a basic assumption. First, the figure of Kidlat's little sister, Alma, enters the film as she is seen traversing her own, small-scale bridge, made up of two overhanging branches, between the village and the outside world (Table 2). Next, she is seen at work, selling ices. She is constructed as the savvy entrepreneur in this scene; unlike the boys, who will suffer from ganging together in a social and competitive group, Alma takes the job seriously and adopts her own, solitary, selling strategy.

This image of Alma as entrepreneur is extended to the next scene where Kidlat is seen to give her a miniature jeepney, in which she can sit and peddle, which replaces the toy she was pulling. The link between childhood and the mobility of the potentially liberating effects of Western entrepreneurialism associated with adulthood are made explicit: 'Alma, look what I brought you. An American gave me dollars. Alma, you are the master of your vehicle. Only you can tell it where to go'. Further, and lest we think the economic power of women is only a future possibility, a woman who is the owner of the ice factory and Ice Drop Shop forebodes and frames the future of Alma. She enters the film as one of Kidlat's passengers, but she is also the object of Kidlat's envy, because she was the first person in the village to ride in an airplane, which is a form of mobility that effectively represents her movement between worlds.

TABLE 2 Alma

Event	Context/setting	Dialogue
Alma crossing makeshift bridge	The bridge is comprised of two branches, one over the other, lying no more than five feet above a small creek. The camera focuses first on Alma's bare feet, moves to her body, and then follows her as she crosses the bridge, walking on the lower branch, hands raised to the higher one to steady herself. A boy is seen using a pole to slowly push a small boat under the bridge only seconds after she passes.	The only thing that is heard is quiet folk music.
Alma preparing to sell ices	Boys and girls come out of the Ice Drop Shop with their boxes of ices, so they may sell them. Alma is the last child who comes out. The camera follows her from behind as she walks down the road, alone, and then shifts to a group of boys laughing in a group, and finally, centres on the face of a very old woman.	Kidlat voiceover: 'My sister Alma, she is smart. She goes her own way. Not like the boys, always competing.'
Alma driving in new toy jeepney	Kidlat has made a small-scale jeepney in the factory alongside construction of full-sized jeepneys; he gives it to Alma to replace her pull-toy version.	Kidlat: 'Alma, look what I brought you. An American gave me dollars. Alma, you are the master of your vehicle. Only you can tell it where to go.'

Public, formal work, therefore, is not simply the preserve of men, nor is it limited to a utopian future wherein women can 'join' men at work. Rather, it is the individualised woman, represented in the form of Alma and the Ice Drop shop owner, who are portrayed as particularly skilled in business. Indeed, their individual pragmatism stands in sharp contrast to the village boys on the one hand, who, as we see later in the film, are to share a ritualistic circumcision as their rite of passage to manhood, and Kidlat himself who must travel to Europe and undergo a series of magic realist practices in order to finally 'grow up'.

Second, although Kidlat's mother is confined to the spaces of the home, it is she who articulates the cutthroat capitalism that lies at the heart of American geopolitics. Returning to Table 1, it is the mother who tells the story of how the Americans came to wield such economic and political influence in the Philippines through the fable of Kidlat's father who, after being promised independence by the Americans in return for fighting the Spanish, is subsequently shot for trespassing on what actually becomes American property, namely the San Juan Bridge. As she tells of the father's fate at the hands of the American military forces, we see a slapstick reconstruction of events. It is Kidlat's mother who invokes a militant resistance to this occupation, as she hands Kidlat the horse she has carved from the butt of his father's rifle. She ends the scene with the comment that 'for twelve million dollars, we became U.S. military property'.

This telling of an economic and political history stands in sharp contrast to another articulated a few moments later, this one by Kidlat's friend, Kaya. In Kaya's tale, Kidlat's father takes on a mythical persona, literally blowing away the American forces with the strength of a typhoon. The time horizon in this story does not work to a simple 'past–present–future' as events become jumbled up together, moving in and out of sequence and leaping back and forth over vast numbers of years – it is what Jameson has referred to in another context as a 'history-with-holes'.[61] This is, though, very much a place-based, as well as character-driven tale, as the typhoon has previously been noted as the force against which the local bamboo huts, as well as the village's bridge, are built and which Kaya himself has learnt to construct. Kaya goes on to suggest that Kidlat take note of his father's example; in a somewhat tortured metaphor, he recommends that: 'when the typhoon blows up its cocoon, the butterfly embraces the sun. The sleeping typhoon must learn to blow again.' This metaphor is somewhat confusing because Kaya himself is associated with the butterfly. As can be seen in Table 3, it is tattooed on his chest, and is cited by Kidlat as initially a threatening design. Later, Kidlat learns that the butterfly on Kaya's chest is a safe shelter from the bombs dropped by American jets. The typhoon that blows up the cocoon, then, is that associated with American military force; in counterpoint, the typhoon associated with Kidlat's father lies sleeping, in so far as his son continues to identify with American

TABLE 3 The Butterfly

Event	Context/setting	Dialogue
Live butterflies	Hundreds of small white butterflies are seen flying around for nearly 20 seconds in a rural setting with no human movement, only the moving butterflies. The camera never wavers. The image is the first following the film's opening action and credits.	Silence
Speaking of Kaya's butterfly tattoo	Kaya and Kidlat are sitting under a tree, where Kaya and others are building a home of bamboo, now a fading art, once practised by Kidlat's grandfather. Kaya's shirt is off, and across his chest and arms is a large and elaborate tattoo of a butterfly.	Kaya: 'Kidlat, one day you will come to know the strength of bamboo.' Kidlat voiceover: 'When I was a child, we were all afraid of Kaya because of his tattoos. One day the first American fighter jets came. I ran, terrified. I found myself in the embrace of the butterfly on his chest. I lost my fear of flying machines and we became friends.' During the conversation the camera shifts back and forth from Kidlat to a focus on Kaya and his tattoo.
Speaking of the typhoon and the butterfly	Kaya and Kidlat are talking under a lean-to, which is under some trees. The conversation is about the San Juan Bridge incident, the blowing of the mythic wind by Kidlat's father, and his death at the hands of the Americans. The father has approached the American sentry, who has refused his father the right to pass over the bridge.	Kaya: 'Your father took a deep breath. He blew with a fury that knocked the gun down. Stronger than the winds of Anok mountain, Kidlat. Fifteen more Americans fell before they finally stopped your father. Kidlat, when the typhoon blows up its cocoon, the butterfly embraces the sun. The cocoon must learn to blow again'. During Kaya's talk the camera focuses closely on the butterfly and slowly pulls away. When he is finished, he stands and puts in on his shirt, covering the butterfly for the first time.

technological innovation and prowess, exemplified by his leadership of the Werner von Braun fan club.

It very much matters, therefore, as to who is telling the history of people and place; for Kidlat's mother, it is a clear-cut case of political conquest and economic dependency, whose key events are mappable over time as well as space with great precision. For Kaya, it is a monumental contest between two powerful forces, each manifest in a variety of mythical forms.

The association between the male protagonists and a mythological unfolding of forces extends to Kidlat himself. Kidlat does not directly reference the destructive aspects of American power: instead, he invokes the spectre of this powerful force through the figure of the white caribou. As can be seen in Table 4., the presence of the white caribou presages the first hearing of the Voice of America. Later, it is the focus of attention while Kaya explains the duplicity of the American forces:

> The white caribou is rare. It is born against nature. The white caribou is beautiful, but inside it is cold and aggressive. One day, Kidlat, you will understand that the beauty of the white caribou is like the sweetness of the chewing gum the America soldiers gave you.

And yet, as can be seen in Table 4, the white caribou is also found at the scene of Kidlat's circumcision, watching with cold eyes his entry into manhood. The caribou is not simply a metaphor for American might, then, but stands in as a place-based reference for something much more general, denoting, perhaps, the loss of innocence as childish practices and dreams are gradually, and sometimes all too abruptly, left behind.

What is happening here, we suggest, is that the economic, political and cultural geographies manifest in the film are being differentially rendered through gender. Kidlat, his father and Kaya are explicitly engaged in a spiritual endeavour, which is articulated though their day to day activities (building bamboo huts that will withstand the power of the typhoon), their episodic crises (fighting US troops with the force of the typhoon and the resilience of the butterfly) and their meditations on these. These are all very much place-based experiences and performances, as the film's male protagonists are tied into the surrounding landscape through their association with mythical figures. Indeed, it is through these place-based identities that the economic and political might of the US, as well as its cultural hegemony, is recognised and mediated.

In contrast, the female protagonists, though ostensibly 'anchored' to home and family through familial ties, are also seen to move in a world of economic and political dependence, wrought from years of American imperialism. They are portrayed as able and willing to adapt to this circumstance, but also as active critics of this same state of affairs, surely fulfilling Jameson's call for a cognitive mapping of one's place within late capitalism. They are even kept apart from the place-based, culturally specific, mythological aspects of the film, such that their identity is firmly tied to this overarching political economic 'reality.'

Though Jameson directs his attention solely to the activities of Kidlat, we can see that there are actually two modes of accommodation and resistance at work in *The Perfumed Nightmare*. Moreover, gender is significant not only in regard to *who* is locating and articulating the 'problem' of Philippine society, it also matters in regard to *how* that problem is to be resolved. As Jameson himself notes in regard to the revolutionary tone of the central male protagonist, Kidlat:

> The beauty of the resolution on this particular level – the way in which the image of a butterfly that enfolds the sun in its wings unites the gentleness of the Kidlat persona with the violence of revolutionary rage itself – is marked as a fragile figure by the very nature of its content. ... the image

TABLE 4 The White Caribou

Event	Context/Setting	Dialogue
First images of the white caribou.	The caribou is seen standing next to a tree, overlooking the village, which occurs within 30 seconds of opening images of butterflies that follow the opening credits. The camera then cuts to Kidlat waking up to Voice of America, then returns to the caribou; after a few seconds, the caribou disappears and only the tree is seen.	During the first image of the caribou, quiet traditional music is heard. When Kidlat is first seen sleeping, a local Filipino broadcast is played, which quickly shifts to the Voice of America. Upon return to the caribou, only the music is heard, then the precise moment the caribou disappears, the Voice of America says: 'Get involved in the Bicentennial…'
The White Caribou tethered.	The close-up image of the caribou is seen right after Kidlat is shown sleeping under a tree, with farm animals wandering around.	There is a voiceover from Voice of America, which begins with the countdown and take-off – while focused on Kidlat – and then shifts to 'One small step for man…' while looking at the close-up of the caribou.
The White Caribou, part 1 of a continuous sequence: moving up a hill.	The caribou moves up a hill on a tether held by a man, moves past the same tree and angle shown in the initial images, and then stops near the top of hill, with boys playing basketball in close background.	Kidlat to Kaya: 'Today is the birthday of Werner von Braun. He's an American immigrant; I didn't learn that in school.' Then, while the caribou continues up the hill, there is just quiet, traditional music.
The White Caribou, part 2 of a continuous sequence: rural setting.	The camera shifts from boys playing basketball to the caribou in a rural setting. Kids go by on a 'normal' caribou; then the camera goes to Kidlat talking with Kaya.	Kidlat to Kaya: 'Ah yes, Kaya, I dreamt once again of the white caribou. Kaya, why is the eye of the white caribou so cold?'
The White Caribou, part 3 of a continuous sequence: "beauty of the caribou".	The caribou is shown with a boy standing in front of him, holding a basketball, then the camera shifts to a close-up of the animal, and then to a close view of its eyes.	Kaya to Kidlat: "The white caribou is rare. It is born against nature. The white caribou is beautiful, but inside it is cold and aggressive. One day, Kidlat, you will understand that the beauty of the white caribou is like the sweetness of the chewing gum the America soldiers gave you."
The White Caribou, part 4 of a continuous sequence: circumcision.	Kidlat and Kaya are walking through the forest, with the caribou in the background as small statue. Then a long circumcision scene begins (apparently years past, or present day, with Kidlat reminiscing) with boys playing in the water, which then shifts to five boys in various stages of crude circumcision in the forest. During Kidlat's circumcision (as a boy), the camera pans down a tree to a hole, where a carved statue of the caribou is seen to be set. The camera then focuses on the statue's eyes, which seem to be richly carved with detailed black lines.	Kidlat to Kaya, as they walk through the forest: "Kaya, I now recall. I first saw the white caribou somewhere here many years ago on the day that I became a man." During Kidlat's circumcision scene, he continues: "While they were hammering at me, I could not take my eyes off a strange leafless tree. The white caribou stared so coldly, I felt no pain."

> itself, which we here manhandle with clumsy fingers, has all the brittle delicacy of the butterfly's wings, no matter what may be the ultimate destiny of this figure – from tattoo to historical force of nature. This is perhaps to say that the cultural-nationalist alternative – a politics which draws on indigenous cultural traditions in order to summon the force and will to dislodge the invader – is here inscribed as an impulse rather than a program, as an aesthetics of revolt rather than its concrete politics.[62]

We would go further here and note that this brittle, fragile figure of resistance stands in sharp contrast to that proffered by Kidlat's mother, namely the butt of his father's rifle. If we take the Westernised view of femininity as delicate, pretty and intransitive, then Kidlat's resolution certainly fits the bill.

CONCLUSIONS

As we hope to have made clear, Jameson's understanding of the spatialities at work in society asserts a hierarchical, systematic view of the world, wherein the economy is presumed to operate at the global level and both culture and politics are relegated to a contingent position at the local level. Because current cultural forms such as film are produced within the totality that is late capitalism, they allow for the suppression of critical knowledge of this prevailing state of affairs (thereby operating as ideology). And yet, these forms also have a utopian dimension, whereby the desire for a better world can be articulated and grasped, albeit largely unconsciously. What makes *The Perfumed Nightmare* meaningful for Jameson, therefore, is not its 'Third Cinema' credentials as a critique of imperialism, but its particular form of cognitive mapping, whereby the key protagonist comes to appreciate the place of the individual within the world-system. The contingencies of culture and politics, therefore, are significant only in so far as they allow Kidlat's unique form of cognitive mapping to take place. Small wonder, then, that gender plays no role in Jameson's analysis of *The Perfumed Nightmare*.

In response, we have noted the logic of equivalence at work in Jameson's broad-scale theorisation of cultural forms, in that Jameson's argument actively subsumes various theorisations on sexuality, gender, race and so on (including some labeled 'postmodern'): such bodies of knowledge have been located, sorted and placed according to the extent to which can further illuminate the project of Hegelian Marxism. This same logic is then projected onto the social realm, as the ontological categories derived from Jameson's understanding of the broad, pervasive, deep set of social relations that is late capitalism – namely, the economic, the cultural and the political – overwrite those others that highlight specificity, variation and heterogeneity. In recovering the geographies of gender present within *The Perfumed Nightmare*, we hope to have highlighted how important it is to pay attention to how such categories are articulated and framed such that they become meaningful within particular contexts.

As we noted in the introduction, such a critique necessarily leads us to question Jameson's dismissal of a supposedly 'progressive' Third World cinema. Recall that for Jameson resistance is very much tied into a critical awareness of one's place within the world system. Because it works to foster archaic nationalistic and fundamentalist tendencies, Jameson argues, Third Cinema simply does not perform a productive function.[63] In so far as Third World cinema is conflated with a state-sponsored 'new cinema', developed as an anti-colonial recovery of an indigenous culture, we would concur with just such a conclusion. Where we do differ is in regard to Jameson's 'fixing' of the form and function of this object of analysis. For Jameson, Third World cinema is *necessarily* all about the national, a condition ensuing from their affectation of Realism as well as their anti-colonial function.[64] And yet, as Bhabha, Spivak and Prakash,[65] amongst others, have argued, it is possible to conceive of cultural forms not as the purveyors of an undifferentiated, textualised subject – variously, First or Third World – but as a venue through which the logic of oppositionality can itself be opposed through the careful deconstruction and re-negotiation of the proffered terms of debate. Thus, while we ourselves cannot commit to Third Cinema as a homogenous entity with known and fixed characteristics, we do see the potential for a productive exploration of this category in so far as films produced and consumed under this heading lay claim to re-present people and place.

This re-envisioning of Third Cinema as an object of analysis can be taken as part and parcel of a broader-based project on the representation of a 'geo-politics', as O'Tuathail has named it.[66] Firmly rooted in anti-essentialist thought and animated by case-specific deconstruction, such a project points to the manner in which the identities of people and place are imagined and re-imagined through the production of expert and popular knowledges, and, moreover, inquires into the question of to whose benefit particular representations are promulgated and maintained. Sharp's thorough analysis of the *Reader's Digest*, for example, highlights the complex linkages between Cold War rhetoric and masculinity, while Dodds and Kirby refer to the proliferation of such state-centred rhetorics through diverse mediums, from cartoons to boys own comics.[67]

And yet, we would hope that this analysis points to something more than another instance of critical geo-politics at work; that is, we wish to emphasise not only how notions of the political, the economic and the cultural, as well as the relations between them, are socially constructed, but also how particular case studies are deployed as exemplars, illustrations or even symptoms within such bodies of knowledge. More often than not, conceptual differences are taken as the grounds for engagement between, say, poststructuralism and historical materialism. Methodological issues tend to be dismissed as the mere nuts and bolts elements of a research project. As we hope to have shown in the case of *The Perfumed Nightmare*, the manner in which objects of analysis (including the subject) are located, identified and assessed in relation to other objects is not simply a product of theory, but is crucial to the way in which theory is imagined and re-imagined. In this instance, *The Perfumed Nightmare* has been re-read as a partial and contingent assemblage of gendered discourses on the character of the economic, the political and the cultural. Accordingly, we can deploy the term 'cognitive mapping' but, rather than have it refer to the imperative to locate the self in relation to the Real, we can rework it as the attempt to outline the web of significations within which objects are embedded and, we would hope, the concomitant lines of fracture and contradiction that allow for such objects to become meaningful in a host of other contexts. A more challenging project to pursue, perhaps, in so far as it implies a continuous process of reworking the subject matter, but one which offers other viewers, public and academic alike, a 'place' in the re-imagining of themselves and the world around them.

NOTES

1. There is no doubt that Jameson's impact on social theory in general has been profound. His work on postmodernity as a 'condition' has become required reading in a number of disciplines, including International Relations (particularly with regard to globalisation and the 'hollowing out' of the nation-state) and geography (as part of the Marxist critique of late capitalism's erasure of place). And yet, his work on cinema has been unevenly received. For some writers on contemporary film (such as Denzin, and Shiel and Fitzmaurice) Jameson's critical account of culture provides a foundational framework for research into film, while some 'definitive' tomes, such as G. Nowell-Smith's *The Oxford History of World Cinema: The Definitive History of Cinema Worldwide* (Oxford: Oxford University Press 1996) bear no reference to his work. See N.K. Denzin, *Images of Postmodern Society: Social Theory and Contemporary Cinema* (London: Sage 1991); M. Shiel and T. Fitzmaurice (eds), *Cinema and the City: Film and Urban Societies in a Global Context* (Oxford: Blackwell 2001). Within the subfield of film geography, Jameson's essays on cinema are notably absent from S. Aitken and L.E. Zonn's edited collection *Place, Power, Situation and Spectacle: A Geography of Film* (Savage, MD: Rowman & Littlefield 1994); and even from G.M. Macdonald's assessment of The Perfumed Nightmare, 'Third Cinema and the Third World', in Aitken and Zonn (note 1) pp.27–45. Instead, film geographers have preferred to reference the work of Marxian cultural and economic geographers, such as D. Cosgrove and S. Daniels (eds), *The Iconography of Landscape* (New York: Cambridge University Press 1988); and D. Harvey, *The Condition of Postmodernity: An Inquiry Into The Origins of Cultural Change* (Oxford: Blackwell 1989) pp.308–26; as well as cultural materialist writers on the media, e.g. R. Williams, Culture and Society, 1780–1950 (New York: Columbia University Press 1958, repr. 1983). Recent anti-essentialist collections such as D. Clarke (ed.), *The Cinematic City* (London: Routledge 1997) and T. Cresswell and D.P. Dixon, 'Engaging Film', in T. Cresswell and D.P. Dixon (eds), *Engaging Film: Geographies of Mobility and Identity* (Lanham, MD: Rowman and Littlefield 2002) pp.1–10, mention Jameson only in passing. What makes this omission something of a shame is that Jameson's work on cinema is so avowedly geographic.

2. See especially Macdonald (note 1); A. Sison, '*Perfumed Nightmare* and Negative Experiences of Contrast: Third Cinema as Filmic Interpretation of Schillebeeckx', *Journal of Religion and Film* 6/1 (2002) pp.1–8.

3. See D.P. Dixon and J.P. Jones III, 'My Dinner with Derrida, Or, Spatial Science and Poststructuralism Do Lunch', *Environment and Planning A* 30 (1998) pp.247–60; idem, 'Poststructuralism', in

J. Duncan, N. Johnson and R. Schein (eds), *A Companion to Cultural Geography* (Oxford: Blackwell 2003) pp.79–107.

4. See T. Barnes, *Logics of Dislocation: Models, Metaphors and Meanings of Economic Space* (New York: Guilford Press 1995); D.P. Dixon and J. Grimes, 'On Capitalism, Masculinity and Whiteness in a Dialectical Landscape: The Case of Tarzan and the Tycoon', *GeoJournal* 59.4(2004) pp.265–75; Dixon and Jones, 'Poststructuralism' (note 3); K.-J. Gibson-Graham, *The End of Capitalism (As We Knew It): A Feminist Critique of Political Economy* (Oxford: Blackwell 1996); L. Hutcheon, *The Politics of Postmodernism: New Accents* (New York: Routledge 1989).

5. First Cinema consists of an industrialized Hollywood film industry, while Second Cinema consists of European art-house films; F.E. Solanas and O. Getino, 'Towards a Third Cinema', in S. Field and P. Sainsbury (eds), *Third World Cinema* (London: Afterimage Publishing 1969) p.29.

6. Ibid.

7. F.E. Solanas, 'L'Influence du Troisième Cinéma dans la Monde', *CinémAction* (1979).

8. See P. Willemen, 'Third Cinema in Question: Notes and Reflections', in J. Pines and P. Willemen (eds), *Questions of Third Cinema* (London: British Film Institute 1989) pp.1–29.

9. T.H. Gabriel, *Third Cinema in the Third World: An Aesthetic of Liberation* (Ann Arbor, MI: University of Michigan Press 1982).

10. Willemen (note 8) p.3.

11. J. Pines and P. Willemen (eds), *Questions of Third Cinema* (London: British Film Institute 1989).

12. See for example M. Wayne, *Political Film: The Dialectics of Third Cinema* (London: Pluto Press 2001).

13. F. Jameson, *Sartre: The Origins of a Style* (New Haven, CT: Yale University Press 1961); idem, *Marxism and Form: Twentieth Century Dialectical Theories of Literature* (Princeton, NJ: Princeton University Press 1971); idem, *The Prison-House of Language: A Critical Account of Structuralism and Russian Formalism* (Princeton, NJ: Princeton University Press 1972).

14. The 'Frankfurt School' refers to a group of philosophers loosely organized around the Frankfurt Institute for Social Research in 1930s Germany. Their radical cultural criticism addressed art, music and film, amongst other media, as key components of social hegemony in an era of state-monopoly capitalism; see M. Jay, *The Dialectical Imagination: A History of the Frankfurt School and the Institute of Social Research 1923–1950* (Berkeley, CA: University of California Press 1996). Though the school is, perhaps, best known for its critique of a 'passive' consumer society socialized by the 'culture industries' (see T.W. Adorno, *Culture Industry: Selected Essays on Mass Culture* (London: Routledge 1991); T.W. Adorno and M. Horkheimer, Dialectic of Enlightenment (New York: Herder and Herder 1972)), Walter Benjamin, 'The Work of Art in the Age of its Mechanical Reproducibility', in W. Benjamin (ed.), *Illuminations* (New York: Harcourt 1955, repr. 1968) pp.219–53 offered a distinctively optimistic take on the utopian potential of emergent cultural mediums, particularly film. Put very briefly, he argued that they created a new kind of spectator, able to critically appraise such cultural forms.

15. F. Jameson, 'On Zardoz', *JumpCut* 3 (September/October 1974).

16. F. Jameson, 'Class and Allegory in Contemporary Mass Culture: *Dog Day Afternoon* as a Political Film', *Screen Education* 30 (1977) pp.75–92.

17. F. Jameson, '*The Shining*', *Social Text*4 (1981) pp.114–25.

18. F. Jameson, 'On Magic Realism in Film', *Critical Inquiry* 12/2 (1986) pp.301–25.

19. F. Jameson, *Signatures of the Visible* (New York: Routledge 1990).

20. These lectures were subsequently published as F. Jameson, *The Geopolitical Aesthetic: Cinema and Space in the World System* (Bloomington, IN: Indiana University Press 1992). In 1986, Jameson attended a conference at Duke University on the 'Challenge of Third World Culture', at which Kidlat Tahimik exhibited some rushes from his *Perfumed Nightmare* project.

21. E. Mandel, *Late Capitalism* (London: New Left Books 1975).

22. F. Jameson, 'Postmodernism, or, The Cultural Logic of Late Capitalism', *New Left Review* 146 (1984) pp.53–92.

23. See F. Jameson, *The Political Unconscious: Narrative as a Socially Symbolic Act* (Ithaca, NY: Cornell University Press 1981).

24. See F. Jameson, *Postmodernism, Or, The Cultural Logic of Late Capitalism* (Durham, NC: Duke University Press 1990), pp.50–54, 409–17; C. MacCabe, 'Preface', in F. Jameson, *The Geopolitical Aesthetic: Cinema and Space in the World System* (Bloomington, IN: Indiana University Press 1992) pp.ix–xvi.

25. Jameson, *The Political Unconscious* (note 23) pp.288–91.

26. Benjamin (note 14)

27. Jameson, *Signatures* (note 19) p.34.

28. F. Jameson, 'Beyond Landscape', in Jameson, *The Geopolitical Aesthetic* (note 24) p.1.

29. N. Deocampo, *Short Film: Emergence of a New Philippine Cinema* (Manila: Communication Foundation for Asia 1985) p.18.

30. Sison (note 2).

31. F. Jameson, '"Art Naïf" and the Admixture of Worlds', in Jameson, *The Geopolitical Aesthetic* (note 24), pp.187–8.

32. Ibid., p.189.

33. Jameson, 'Beyond Landscape' (note 28) p.191.

34. Jameson, '"Art Naïf"' (note 31) p.199.

35. Jameson, 'Beyond Landscape' (note 28) pp.197–8.

36. Jameson makes the interesting observation that such a representation of space deliberately undercuts the assumption one might hold that modernization diffuses outward from the First to the Third World (ibid. p.204).

37. Jameson, 'Postmodernism' (note 22); Jameson, *Postmodernism* (note 24).

38. Jameson, 'Beyond Landscape' (note 28) p.197.

39. Ibid. p.190.

40. Ibid. pp.194–5.

41. Ibid. p.201.

42. There is also a Third World audience who, Jameson suggests, will appreciate Kidlat Tahimik's 'naïf' use of Super 8mm film, because it symbolises his refusal to be thoroughly co-opted into commercial cinema; Jameson, '"Art Naïf"' (note 31) p.203.

43. Jameson, 'Beyond Landscape' (note 28) p.202.

44. Ibid. p.210.

45. Spain dominated the region from1565 to 1898, whereupon the US 'bought' the country as a result of its victory in the Spanish American War in 1898, thus providing a significant political and cultural shift: 'to paraphrase Nick Joaquin, the preeminent Philippine fiction writer, after more than 350 years in a convent, it was time for Hollywood' L. Francia, 'Philippine Cinema: The Struggle Against Repression', in J.D.H. Downing (ed.), *Film and Politics in the Third World* (New York: Praeger Publishers 1987) p.209. Following the Japanese occupation of 1942–1945, the United States relinquished its hold, resulting in a formally independent Philippines in 1946.

46. Jameson, 'Beyond Landscape' (note 28) p.211

47. See R.B. Tolentino, 'Jameson and Kidlat Tahimik', *Philippine Studies* 44 (1996) pp.113–25.

48. See Deocampo (note 29); Francia (note 45); D. Joel, 'Philippine Film Industry as Postcolonial Discourse', in R.B. Tolentino (ed.), *Geopolitics of the Visible: Essays on Philippine Film Cultures* (Quezon City, Philippines: Ateneo de Manila Press 2000) pp.3–9; N.C. Tiongson, 'The Filipino Film Industry', *East-West Film Journal* 6/2 (1992) pp.2361; and R.B. Tolentino, 'Introduction', in Tolentino, *Geopolitics of the Visible*, pp.vii–xxi.

49. F.C. Lim, '*Perfumed Nightmare* and the Peril's of Jameson's New Political Culture', *Philippine Cultural Forum* 1/1 (1995) pp.24–37.

50. Tolentino, 'Jameson' (note 47).

51. E. San Juan Jr., 'Cinema of the "Naïve" Subaltern in Search of an Audience', in Tolentino, *Geopolitics of the Visible* (note 48) pp.269–70.

52. Ibid., p.271.

53. D. Harvey, *Justice, Nature and the Geography of Difference* (Cambridge, MA: Blackwell 1996).

54. Tolentino, 'Introduction' (note 48).

55. Ibid. p.vii.

56. Jameson, *Postmodernism* (note 24).

57. Ibid. p.6.

58. See especially E. Soja, *Postmodern Geographies: The Reassertion of Space in Critical Social Theory* (London: Verso 1989).

59. Hutcheon (note 4) p.17; see also Deutsche's critique of the work of David Harvey; R. Deutche, 'Boys Town', *Environment and Planning D: Society and Space* 9/1 (1991) pp.5–30.

60. Gibson-Graham (note 4) p.121.

61. Jameson, 'On Magic Realism' (note 18).

62. Jameson, '"Art Naïf"' (note 31) p.208.

63. Jameson, *The Geopolitical Aesthetic* (note 24) pp.187–8.

64. Jameson, 'On Magic Realism' (note 18); F. Jameson, 'Third World Literature in the Age of Multinational Capital', *Social Text* 15 (1986) pp.65–88.

65. H. Bhabha, 'The Commitment to Theory', in J. Pines and P. Willemen (eds), *Questions of Third Cinema* (London: British Film Institute 1989) pp.111–132; G.C. Spivak, *In Other Worlds: Essays in Cultural Politics* (New York: Methuen 1987); G. Prakash, 'Writing Post-Orientalist Histories of the Third World: Perspectives from Indian Historiography', *Comparative Studies in Society and History* 32 (1990) pp.383–408.

66. G. Ó Tuathil, *Critical Geopolitics* (Minneapolis MN: University of Minnesota Press 1996).

67. J. Sharp. 'Reel Geographies of the New World Order: Patriotism, Masculinity, and Gepolitics in Post-Cold War American Movies', in G. O Tuathail and S Dalby (eds), *Rethinking Geopolitics* (London: Routledge 1998) pp.152–69; K. Dodds, 'Reel Geopolitics', *Media Education Journal* (Autumn 2003) pp.3–8; A. Kirby. 'The Construction of Geopolitical Images: The World According to Biggles (and other Fictional Characters)', in K. Dodds and D. Atkinson (eds), *Geopolitical Traditions* (London: Routledge 2000).

Palestinian 'Roadblock Movies'

NURITH GERTZ
Tel Aviv University, Tel Aviv, Israel

GEORGE KHLEIFI
Institute of Modern Media, Al-Quds University, Jerusalem, Israel

The beginning of Palestinian cinema as an art form can be traced back to Michel Khleifi's first film *Fertile Memories*.[1] In the 1960s and1970s, the years that preceded this film, the Palestinian cinema was mainly a documentary one, operating under the wings of the Palestinian organisations and directly partaking in consolidating the Palestinian nation and its historical narrative. For years, the Palestinian society had been composed of regional, local, familial, extended-family affiliated, religious and pan-Arabic identities competing with the national identity.[2] Khalidi[3] illustrated this by stating that a Palestinian from Nablus, for example, had defined himself first and foremost as a Nablusi or as an Arabic person and only later as a Palestinian.

With the rise of Palestinian nationality from the beginning of the twentieth century and particularly since 1948, the Palestinian discourse, and within it the Palestinian cinema, have been striving to relate these identities to nationhood and, thus, to incorporate them into a national identity. Lacking a homogeneous state sprawling over a homogeneous space,[4] the Palestinian discourse and cinema have nevertheless attempted to create a homogeneous historical narrative.

Various historians argue that it is precisely the experience of being uprooted, the refugee condition and the disappearance of the local identity – on the one hand – and the isolation of the Palestinians in the various hosting societies – on the other – that have accelerated the consolidation of this narrative. The national consciousness was strengthened at the expense of the regional and communal identities that have lost their power and significance.

Nassar[5] mentions the important role that literature, and especially poetry, played in this process.[6] The Palestinian cinema of the 1960s and 1970s was meant to play a similar role.

The common story shaped by the documentary cinema during those years filled the function of constructing a collective identity in the place where the old institutions constructing such identities in the past (the family, the religion, the village, the region) disintegrated.[7] It was shaped at the historical moment in which the need to replace the different identity narratives with a comprehensive, unifying identity one arose. Not the separate villages and families were at the center of attention anymore, but those who had originated from these villages and families. The new narrative would reintegrate what had disintegrated. The need of that cinema 'to legitimize and strengthen national identity' has led it to create an 'uncomplicated, linear national history'.[8] It invents what Said calls 'the rhetoric of belonging',[9] and attempts to 'create out of the fragmented history of exile a new unity'.[10] Hobsbawm[11] describes the invention of tradition as a praxis serving the authorities in mass societies to create national bonds as a replacement of the connections within the small unites, such as the village and family that have crumbled.[12] The 'cinematic' Palestinian story in the 1960s and 1970s does not strive to reconcile and bridge between the private, divided stories, since it does not recognize, in fact, privacy or division. Instead, it strives to incorporate these stories into a linear, comprehensive and exclusive one. And yet, it seems that in fact it fills the same function as 'the invention of tradition' does, according to Said[13] and Hobsbawm.[14] This narrative was constructed at that period as a homogeneous story, from which the Palestinian nation evolves by creating a relationship with the outside against which they define themselves.

Khleifi's film *Fertile Memories* was the first important film substituting the Palestinian 'time story' for a story of 'space'. It testifies as to the process, described by Yiftachel,[15] in which 'territory (the where of the nation) becomes a most vital "kernel" of national mobilization, while the history of national origins (the "when") tends to become mythical and homogeneous, used chiefly to boost the territorial struggle'.[16]

However while Michel Khleifi's films reflected the process in which territory became the main – though by no means sole – shaper of the nation,[17] they also deconstructed this territory in order to reveal behind it the multiplicity of the Palestinian society.

Michel Khleifi's first film, *Fertile Memories*, and others that followed have two facets. While they depict the local space, centered on and around, the family home and the village or town in which the heroes' daily lives come to pass,[18] they also portray a national place that encompasses the horizon of what is considered 'the Palestinian space'. On the one hand this space is divided between families, villages, genders and classes, while on the other it is unified around memories of a common, national past, the one that preceded the war of 1948 and the creation of the State of Israel. This past serves as an object that although lost, continues to exist in the films through the depicted space.

Khleifi's films were created in the context of a Palestinian society in which regional, familial, religious or pan-Arabic identities vie with the national identity. These films restore their independence alongside the national one, and in doing so construct an open, heterogeneous unity, a national and a third space together.[19] They present, as Shohat and Stam[20] contest, the differences, gaps, and tensions within Palestinian society, but also reflect a common struggle against the occupation, and a shared cultural, historical identity, whose roots are in the land and in the past.[21]

Khleifi is considered the founder of the new Palestinian cinema, as well as one of its greatest directors. But the same kind of harmony depicted in his films was also evoked by a number of other Palestinian directors – most of whom were born and raised in the villages and towns of their forefathers and that are now within the borders of the State of Israel. Given their background, they were able to integrate the landscape of memory and the landscape of the present. Among them are Ali Nassar,[22] Nizar Hassan,[23] and Hanna Elias.[24] These directors and others, especially those who were exposed to several cultures in Europe, the United States or Israel were more prone to also deconstruct this landscape.

In contrast, the Palestinian artists who began to direct films in the 1990s, had grown up under Israeli occupation, mainly in the West Bank, in refugee camps, in Gaza, and matured or worked in this decade of increasing national tension during and between the two *intifadas*. These filmmakers have found it difficult to create the spaces that Khleifi, Elias, Nassar or Hassan succeeded in creating in their films. They also found it difficult to deconstruct this space as did Khleifi, Elia Suleiman or Rashid Masharawi. As the *intifada* strengthened and the Israeli occupation became more deeply felt, the entire space seemed to be threatened, narrowed or reduced, and the films found it difficult to encompass the wholeness and unity of the home and the outside, to describe the mundane life within them, and to connect them to past times. Many of them also failed to preserve the ambivalence and heterogeneity that made it possible, in Khleifi's, Suleiman's or Masharawi's films, to create the national story from – and through – individual stories, and to connect the private place to the public one. Since the end of the 1990s, and now, at the beginning of the present decade, it has become difficult for the Palestinian cinema both to create and to deconstruct a whole imaginary map – and to produce from this map a heterogeneous, multi-cultural space. In the films made in the 1990s and 2000s the public space is blocked, the private space is missing or destroyed, and the only place left intact is the border.

In this case, however, the border is not the defining line that might divide, as Tuathail and Dalby[25] phrase it, us from them.[26] It does not part any longer the world into good and evil[27] and does not leave the demonic effects out.[28] On the other hand, the border is not a threshold linking between cultural identities.[29]

During these years crossing borders and going through checkpoints have become grueling daily experiences, the number of by-pass roads and settlements cutting through the space of the West Bank has increased, and Israeli soldiers and tanks have intruded time and again into Palestinian cities, streets and houses. One result is that the borders have become emblematic of Palestinian space and identity, which are either dispersed in Israel itself, in the West Bank and the Gaza Strip, in other countries in the Middle East, and all over the rest of the world, or imprisoned within refugee camps and beyond fences and watchtowers, and further divided by numerous Israeli settlements and pass roads. The borders have thus become a sign of oppression characterised by an Israeli definition of Palestinian as a non-existent, split or broken identity. This, perhaps, may explain why so many new Palestinian films take place at borders and checkpoints, and have therefore been termed 'roadblock movies'.

The Palestinian films of the 1990s find it difficult to present 'visual order of space'.[30] They find it complicated to map the Palestinian territory with the aid of the camera or by editing, to clearly distinguish between the inside and the outside.[31] On the other hand, in this situation of an insecure territory, they do not allow themselves to describe 'plurality of space and the multiplicity of possible political constructions of space'.[32] They cannot create maps that put the dispersed pieces together to form a 'larger picture and to draw out connections and causalities between them'.[33] They cannot present the same complicity of constructing and deconstructing that characterised Khleifi's films, to a great extend they revived the cinema of the 1960s and 1970s–a cinema attempting to create a coherent national identity, to strengthen it and legitimise it in the face of the other outside.[34]

In Hani Abu Assad's film *Rana's Wedding*,[35] home is the place from which the protagonist flees and escapes. Rana has been ordered by her father to choose a suitable husband from the list of candidates that he has prepared for her. Her father allows her until four o'clock in the afternoon to decide; if she doesn't do this, she will have to go with him to Egypt. The film's protagonist, Rana, lives in a grand and beautiful home, but throughout the film we see her, and others, running away from it. At the beginning of the film Rana leaves it before daybreak to search for her beloved; when her father agrees to approve the marriage, stricken with fear she flees to the desert that lies just to the east of Jerusalem; and at the end of the film all those invited to the wedding are forced to leave the home and cross the roadblock, where the wedding ceremony is performed. The home thus loses its stability and reliability as a safe haven, protective and warm. Rana's marriage is meant to allow her to build a new home. However, a menacing shadow is seen to hover over this future home as well when she visits a friend and, through the window, witnesses an Israeli bulldozer destroying a neighbor's home: 'They are destroying a home on the very day that I want to build one,' she says.

Set against the home the film presents, through journeys and movements, the outside landscapes of the country. It is a film about journeys: the first is Rana's journey in search of the man she is in love with; then there is a journey Rana and her now found beloved take to find someone to register their marriage; finally, there is a journey back to Jerusalem, in order to convince Rana's father to approve of their marriage. The journeys in *Rana's Wedding*, like those in many other Palestinian films, appear at first to denote mobility, control of the space, the mapping out of the landscapes, and movement towards a goal. But like the journeys in other Palestinian films, this one, too, leads through a destroyed country and ends in only one destination – the roadblock. The city of Jerusalem, where the heroine wanders before daybreak searching for her beloved, is an empty place, frightening and threatening. The places she passes, and those she later travels through with her husband-to-be, are all destroyed places, filled with ruins and piles of dirt, and it appears as if they are traveling round and around in a circle. Here too, as in other Palestinian films such as Hanni Abu Assad's *Ford Transit*,[37] Subhi A-Zubeidi's *Crossing Kalandia*,[38] or Rashid Masharawi's *Ticket to Jerusalem*,[39] the beginning, the mid-point and the final point of each of the journeys is the same – a roadblock – around which the protagonists (and sometimes the camera as well) gather, in endless and inescapable circles. The film ends at the A-Ram roadblock, close to Jerusalem, as the soldiers refuse to let the couple's lawyer cross the barrier and he has to wed them right there. Thus as the home is no longer a safe refuge, and the landscape is in ruins, the most meaningful place in the film is the roadblock; it separates lovers, comes between family members who need to meet each other, separates the different sections of Jerusalem, and even divides houses into separate spaces.[40]

In many films a sense of a lost space is provided by cinematic means and is connected with expressions of static, immobile lives. Tawfik Abu Wael's film *Waiting for Sallah A-Din*[41] presents five separate episodes depicting Palestinian life in Jerusalem: police harassment of street peddlers, the endless queue in front of the offices of the Israeli Ministry of the Interior on East Jerusalem's Abu Shidad Street, an unemployed labourer looking for work, the life of a family whose house has been demolished several times, and the story of a cafeteria worker in the well-known Palestinian theater Al Khakaiati who is also an amateur director. These episodes, rather than dealing with heroic events and great tragedies, focus on the small miseries of daily life. Time stands still in the film, leading nowhere, trudging again and again over the same ground, going through purposeless motions: the endless search for work, houses repeatedly demolished, Sisyphean efforts to maintain fruit stands by the city's walls in the face of police who all too repeatedly confiscate their merchandise.

The daily life that is 'stuck' is connected here to the spaces that are 'blocked'.[42] The camera squeezes between demolished walls; it observes the protagonists through railings, closed doors and opaque windows; it photographs from odd angles and deep focuses that give space the appearance of emptiness and depression. The blocking of the spaces, as well as the unremitting continuation of misery, is translated into cinematic language by the lethargic and static movements of the camera as, for example, it moves slowly along the ever growing queue at the Ministry of the Interior on Abu Shidad Street, rests motionlessly on the face of the man whose house has been demolished, or focuses on the man who sits without moving at the entrance to the Palestinian National Theatre (Al = Hakawati), theatre, waiting for the play to end. The camera's lack of motion is intensified by long takes and the immobility of the people.

A film that made strong connections between the fracturing of space, the loss of home, and 'being stuck at the border' is Rashid Masharawi's documentary *Stress*.[43] Made two years before the outbreak of the second *intifada*, it seems to have foreseen it. It opens with a long shot of the Erez roadblock, through which the residents of the Gaza Strip must pass to get to work in Israel, and ends in another long shot, this time of a Palestinian refugee camp in Gaza. The camera, however, rather than lingering on the spaces captured in the long shots, 'gets stuck' at borders and barriers: different areas within the Israeli checkpoint, at roadblocks and transit points, and at the internal borders that cut and split up the Gaza Strip, and the Palestinian society within it, into tiny pieces divided by Israeli settlements, settlers and soldiers.

The entire film takes place out of doors, with not even one scene shot inside a home. Thus, the 'safe havens' depicted in the films of Khleifi, and of other directors,[44] are not just destroyed but are nonexistent. Yet, the outside, too, seems to be nonexistent. It is broken up by the camera's use of close-ups, or by medium- or high-angle shots that capture only fragmented images – a face or hand, a bag or identity card, a stomach, face or shirt – distorted and disconnected pictures without any establishing shot – that is, without a unifying theme. In addition, the outside is cut off by images of soldiers, settlers and settlements that, by dissolves or sharp editing, replace the shots of refugee camps and Arab cities. In this way the film deals with a disappearing space that is dissected and crushed by borders. As in Abu Wahl's film, here too the crushing of space is connected with the crushing of time. The film seems to be describing ordinary Palestinian life – construction workers and labourers as they work or pass through the roadblocks, children playing etc. Yet the daily life shown here lacks any continuity because each act is cut off from its context and is thus transformed into an isolated event – a moment in time with neither past nor future, and without an all-encompassing life that gives meaning to it.

In all these new Palestinian films the loss of outer space is connected with the loss of the private one. In Masharawi's film the home is missing, and in Azzaa Al Khassan's documentary film *Three Centimeters Less*,[45] the home is destroyed. In the latter, the film turns its back on the landscape and concentrates on homes – on the interiors of apartments, and on the kitchens, living rooms and balconies of the protagonists' houses. However these homes do not serve as a safe alternative to the outside space: 'This is the home', The director says in the opening scene as she moves with her camera in the midst of ruins, 'and this is the attempt to preserve a normal life', she continues as she focuses her camera on a group of children wandering about the ruins. Her film opens with these ruins and ends with the wreckage of her apartment when she returns to it and moves through the rooms and surveys the damage done by the soldiers who have broken into and searched it. The destruction at the beginning and at the end of the film and the disappearance of the spaces throughout, symbolise the impossibility of creating, cinematically, a complete harmonious space.

Other films too relate ironically to the lost private space. In Nizar Hassan's *Invasion*,[46] made when the results of the Israeli invasion of Jenin in 2002 were fresh, children are photographed playing in the rooms of an apartment of which only the walls remain, while a couple sits on the floor on the second story of a half-collapsed house, elegant curtains still billowing in the windows of the destroyed room. The strong effect of destruction in this film is created by the vivid contradiction between the destroyed buildings and the simple acts of daily life, a life that perseveres as if the buildings were still standing intact.

An examination of the differences between Elia Suleiman's 1996 *Chronicle of a Disappearance*[47] and his 2002 *Divine Intervention* can provide a quite thorough summary of how the escalation of the Israeli–Palestinian conflict and the strengthening of the Israeli occupation regime changed the landscapes of the Palestinian cinema. The two films have the same structure. In both of them some of the events take place in Nazareth, in the home of Suleiman's parents, and others in Jerusalem. In both the filmmaker, who adopts a comical stance to his surroundings, makes fun of the daily friction between the Arabs of Nazareth as well as of the Israeli violence in Jerusalem. Both reach their climax in a fantasy of revenge against the Israelis.

An analysis of the difference between the two structurally similar films can clarify the geographical, historical and cinematic changes that took place in the few years that elapsed between them. In the first film there are several open or vast spaces: Suleiman's parents' home in Nazareth, his home in Jerusalem, and the broad expanses of the country seen during travels between them. Although these spaces are indeed stifled and threatened, they can allow for a daily, albeit violence-ridden, routine to take place. The later film explicates what was only implicit in the first film. It no longer has any broad spaces; the house in Nazareth is reduced to the size of its kitchen, with the living room seen only once, when its contents are confiscated by internal revenue deputies; and Jerusalem ends at the a-Ram roadblock. Daily life has been transformed into a static routine ridden with hate, rage and friction.

Palestinian cinema depicts a 'state of siege', but also rebels against it. Because the sense of blocked spaces is so strong in this cinema that it cannot inscribe the map that in Khleifi's films combined home and fields – the private space and the public space –it uses the camera in order to 'capture the heights' and, in this way, to control the space below by its gaze.[48]

Throughout Tawfik Abu Wael's entire film *Waiting for Sallah A Din* there is a hidden struggle between Israelis and Palestinians for the control of space. From the beginning of the film the Israelis are in the high places: Israeli soldiers patrol along the top of the walls of the old city of Jerusalem and over the al-Aqsa Mosque, and Israeli policemen ride on horseback over the heads of the peddlers in the markets, and so on. Although Abu Wael's camera, like the protagonists in the film, is in an inferior position, it doesn't give up. By using such camera techniques as tilt-up, zoom-out and long shot, the soldiers on the city wall disappear, are reduced in size or are lost in the distance or in spaces crowded with Palestinians. The spectacular movements of the camera circumvent what the Israelis see, but show what they perhaps fail to acknowledge. The camera moves upwards, over the heads of the Israelis, or drops low, below their viewing range. It takes in the crowds at prayer in the area surrounding the al-Aqsa Mosque, tracks lines of soldiers, rises again to the *muezzin* in a minaret, and then rises even further up, even to the moon. The camera's wide and slow movements capture the sky and suffuse the protagonists' static and limited space with dynamics and height.

Abu Wael's camera thus compensated for the blocking of space by 'controlling' the air and the sky. In many instances camera shots of this kind allow the directors to film in the places that the Israeli Army usually forbids filming. Yet they have a clear function beyond that: 'The Jew is always high up, in the place of the one in charge', says Subhi Al Zubaidi. 'I create a hierarchy that leads from the ground below to what is above it and above the Israeli'.[49] Hani Abu Assad explains this point even more clearly:

> In order to dictate your reality you must film it from above. You cannot change the reality; you can only overcome it. The crane allows me to go over the border that the officer decides for me at the roadblock and to film beyond it. Cinematic means enable you to cross borders.[50]

Palestinian cinema uses its plots as well in order to overcome siege and borders, in order to 'provide' a coherent 'cognitive map'.[51] The structure of Abu Wael's film *Waiting for Sallah A-Din* 'compensates' for the lost space by using parallel plots, all of them dealing with similar hardships and distress, and also by depicting the exchange of messages broadcast from Palestinian relatives all over the world, who communicate by radio as they have no other means of contact. These greetings and parallel plots bridge the distances and create a single, unified, national fate and identity. A palpable oneness is created by the construction of a homogeneous space controlled from above with the mosque, symbolising its unity, at its centre, and by the plots that connect the Palestinian people with their shared fate, and 'gather them together' from all over the world.

In May Mari's film *Borders of Dreams and Fears*52 the plot charts a new spatial cartography over a scarred landscape, but does this allegorically. The film follows the lives of children in two separate refugee camps: Chatila, in Lebanon, and Dheisha, in the West Bank. The map that the film charts is one of exile and separation, a map determined by the outcome of the 1948 war. Indeed the film's opening shows arrows marking the route taken by refugees fleeing Palestine northwards to Chatila on the one hand, and eastwards to the West Bank on the other. The plot, which delineates the children's daily lives, tends to draw an alternative map that will create links and bridges over the distances imposed by the existing map, and thus resurrect an imaginary Palestinian community.

The route of the letters and videos the children exchange between the two camps marks the first path charted on the map. It links their different stories. The two sites are then drawn closer geographically, as the children come to either side of the Israeli–Lebanese border to meet each other. Next, history leaves its imprint on this geographical chart as the child from Deheishe visits Saforie, from which the family of the girl from Chatila fled in 1948, and sends her a sachet of earth from the village. Eventually the children validate the map seen as the film begins, but this time with the arrows pointing in the opposite directions: from exile inwards, into the lost country. This map operates on several symbolic planes for the film's protagonists. It frees them from the reality of the congested camps in which they live into the vast expanses of space, in particular that of their lost homeland and, in addition, introduces a dynamic of movement into an otherwise stagnant existence.

The protagonists of Masharawi's *A Ticket to Jerusalem* and those of A-Zubaidi's *Crossing Qualandia* overcome both border and roadblock with the help of poetry readings and artistic events they orchestrate on both sides of the fence. In Suleiman's *Divine Intervention* and Abu Assad's *Rana's Wedding* love and marriage cross the borders. Since Suleiman (who is also the protagonist of his film) resides in Jerusalem and his girlfriend in Ramallah they can meet only at the roadblock between the two places, where they hold hands and look at each other – as well as at the events that unfold at the barrier. With the cinematic plots at their disposal, therefore, these films strive to overcome fragmentation and disintegration, to 'steal the borders' and construct a unified national identity and territory.

Many Palestinian films, rather than 'rebelling' against the continual shrinking of space, chose to focus on one small, isolated detail and to turn it into a symbol of wholeness and totality – a synecdochic substitute for the broad landscapes that filled Khleifi's films.

Anton Shammas[53] articulates this as follows: 'For A's grandmother, an old Lebanese refugee, Palestine means nothing more than the lemon tree in the backyard of the house she left in Yaffo, or Yaffa, as she calls it.' The small details thus revive Palestinian history in its entirety, infuse the memory of the pre-1948 past into the present, articulate its frailty, and constitute the dream of its return. These are the concrete details that were meant to revive the place, to re-territorialise it, to present the Palestinian existence in the landscape in face of its concealment by the Israelis.[54] Thus, they become the fetishes that stand for the homeland in general, its expanses, past and future.[55] They become alternatives to national existences that were lost, distanced or gone. They construct what Jameson[56] refers to as a national allegory.[57] They represent the vanished events, or the trauma of having them obscured, in Mitchell's words.[58]

Thus, instead of the land, of which the olive tree was only a small part, as in the films of Khleifi, Nassar or Hassan,[59] the films now focus on the tree itself, its trunk, a leaf, a branch. Rather than filming an entire house within its surroundings, the camera pauses by one detail of the house: an arched doorway, a painted ceiling, a crack in a stone wall. Continuity, memories and dreams are represented by children, and hope is delineated as flower buds, flags and kites, and similar symbols.[60]

Above and beyond all of these, the Palestinian national identity, history and territory are symbolised by the al-Aqsa mosque (Haram a-Sharif), with the gold Dome of the Rock. This ubiquitous symbol, accompanied by the amplified sounds of prayer emanating from the minarets, which are heard by the films' protagonists as they roam between heaps of debris and through the roadblock bypasses,[61] serves as a sacred icon.[62] Its existence as a geographical axis (in the centre of the country) and cinematic focus (at the centre of the frame), its lofty position (emphasised by the angle of the camera) and its dominance over the soundtrack (the *muezzin*'s call drowns out all other sounds), places it as a symbol for Palestinian unity.[63] It serves as a substitute for everything that previously symbolised the Palestinian people and united them over time and space: vast landscapes, fields and homes, the private and the public.

Liana Bader's film *Zaytounat* exemplifies the ways in which the particular – the eponymous olive tree in this instance – can represent space in its entirety and how this image then becomes a unifying symbol. In Palestinian society the olive tree has long been emblematic of 'rootedness, identity and resistance'.[64] Liana Bader uses it as focal point around which four stories revolve. One is the story of an artist who had left the country as a young child and returns to paint its olive trees. In another, a woman who lives in a large stone house, surrounded by a garden in which there are two olive trees, fears for the fate of the house, as well as her own. The other stories are about simple villagers picking the olives from their trees, and who have to make do with the handful of trees left to them after some of their land has been expropriated or some of their trees have been uprooted by Israeli settlers. On the face of it, the different stories construct a heterogeneous stratified society, depicting the real, daily lives of at least some of the characters, whose subsistence is based on the produce of their land. However, the real point of the film is not this portrait of men and women of different classes, ages and locales, but rather their collective story, which is signified by the olive tree. This tree is the symbol for the nation in its entirety – its soil, its rural heritage, and its past and future – and the stronger the threat to all of these becomes, so too does the potency of the symbol. There is great variety in the shots of olive trees, and in how they are cut or dissolved into one another: a venerable gnarled olive tree; a newly planted olive shoot; an uprooted olive tree, vandalised by Israeli settlers; a painted olive tree, standing for the dream of the lost homeland; an olive tree as a postcard – and olive trees as the source of livelihood of those who live on the land. Like the curved arches and coloured floor tiles that represent the typical Arab house, the olive tree is the single object that serves as a symbol for the lost but longed for homeland, its past and future, its memories and reveries, its existence and its threatened fate. It bridges over the distances and differences the film draws between its various protagonists and between various areas of life, and erases the differences between men and women, children and adults, rich and poor. They all become part of the nation embodied by the olive trees.[65]

As symbols became the main way of expressing Palestinian identity many recent films use them as the axis of their plots: they start with cars breaking down, stopping, losing the way, or getting stuck in roadblocks, but they end with kites, flags, birds or horses crossing borders and reaching distances and heights,[66] thus indicating that the power of imagination, dreams and longing can reclaim the blocked national space.

The difference between notions of reality and space in Khleifi's films and in the films of the new cinema can be traced to the use of these symbols. The child protagonist of Khleifi's *The Tale of the Three Jewels*[67] captures birds and takes care of them. Both the captive birds and birds that fly freely in the sky symbolise a free, poetic world the child yearns for. Their symbolic role, however, is blurred by the realistic details of capturing and tending of them. Similarly, a rogue mare that escaped to the fields in the middle of a wedding ceremony in *Wedding in Galilee*[68] served as a complex symbol of a feminine world (shots of the renegade mare are cross-edited into shots of tender, semi-erotic interactions between women), of a national nostalgia, and of universal longings.

Khleifi's camera in general, even when focusing on the allegorical parts of a frame, wove them into the space of country, village and home. The symbolic elements of this space became a focal point for a multiplicity of open meanings, of which the national connotation, although central, was only one. The new cinema strips the details of their plain, realistic meaning and assigns them an unequivocal allegorical significance. The galloping horse in *Light at the End of the Tunnel*[69] or in *A Ticket to Jerusalem*[70] and the bird in the sky in *Borders of Dreams and Fears*[71] or in *Fadwa*[72] do not exist in any real contexts, nor do they represent a range of meanings; rather, they epitomise dreams and national nostalgia, and bring together all those who dream and yearn through them.

Recent Palestinian cinema thus uses details of daily life and landscape as allegorical representations of threatened nationality. All too often, however, the detail has lost the meaning it was meant to represent. Shahada[73] aptly describes this:

> I found myself looking at an olive tree, and as I look at it, it transforms itself, before my eyes, into a symbol … of our struggle, of our loss. And at that precise moment, I am robbed of the tree, and left in its stead with a hollow void into which rage and pain flow.

Some of the symbols forged in Palestinian culture were adopted by television networks in the world. They were added to the images of fighters, of children throwing stones and of destroyed houses, all of which served as symbols for Palestinian history and identity, while at the same time 'minimising'[74] and extinguishing them. This gradual process is explained by Sanbar:

> In the beginning, you were required to burn our features into your gaze, so that you cannot forget our faces, or crease them, or whiten them out. But, in fact, the face became an image and] once you grasp it, the face disappears. Soon the image of the *fedayun* became abstract; concealing the image of the society behind him. After becoming a group of refugees, Palestinians were perceived by those who finally saw them as a group of armed fighters. But who is the one hiding beneath the *kaffiyeh* … The competitors who could not banish the Palestinians once more into the void, tried at least to tie them into a new abstraction: that of the mask-wearing man, the terrorist with a rifle.[75]

As previously shown, this abstraction was also fabricated by the Palestinians themselves.

The problem that faced recent Palestinian cinema, therefore, was how to tell the story of the Palestinians – when the only available tools were an array of fixed images that defined Palestinian society for the west, and a collection of symbols through which Palestinian society consolidates its unity and narrative. Michel Khleifi[76] describes this as an 'automatic' discourse of sorts, a 'language of ready-to-serve images'. Or as Elia Suleiman asked in his *Arab Dream*:[77] 'How can you make a film when all you have are two characters: a victim and an aggressor?' – and then showed that this was possible. However, in order to do this he turned the question itself into the theme of his films in which, by exposing the distance created between symbol and referent and between signifier and signified, he gave the images and symbols new meanings.

Suleiman's film *Divine Intervention* for instance, avails itself of the entire inventory of symbols used by Palestinian culture in general and by Palestinian cinema (including Suleiman's other films) in particular: *keffiye*, rifle, Kalachnikov, hand grenade, broken-down car, barricaded road, the mosque in Jerusalem, pictures of Arafat, kites and birds in the sky, the Muslim symbol of star and crescent, and a map of Palestine. There is hardly a symbol that has ever adorned a Palestinian film or literary work that didn't find its way into this film. However, as in Suleiman's previous films, *Divine Intervention* uses parody, humour and the absurd to reveal the fictitious nature of these images, and presents them as a 'simulacrum that never had an original',[78] or as a stereotype that conceals the impossibility of grasping reality itself,[79] Nevertheless, the film does not fail to seek the truths beyond the symbols and to restore the meanings they have lost.[80] In this respect, and despite the abundance of post = modern techniques employed in his films, Suleiman is not a post = modern director.

The balloon flown by Suleiman over the roadblock and over Jerusalem, bearing Arafat's wrinkled face and over-stretched smile, exemplifies the dual role these symbols play in these films. Although Arafat's smile is caricatured, the balloon carries it over the roadblock, leaving the Israeli soldiers there dazed and baffled, crosses the border, flies in the sky over Jerusalem and over churches and synagogues, and finally comes to rest in the mosque. The accompanying music adds a magical, dreamlike dimension to the sequence. Everything represented in Palestinian cinema by kites, flags, birds, rising to heights, the mosque and the city of Jerusalem, is encapsulated in this flight. The balloon crosses borders, allowing hopes, dreams and the imagination to run free, and it captures in the course of its flight both city of Jerusalem and the mosque that, as noted above, are today emblematic of Palestinian unity. At a press conference in Ramallah, in 2002, Suleiman said, 'The soldier in the roadblock can catch me, but he cannot capture my imagination ... You don't need an identity card or a passport to cross the border in your imagination.' Even though the balloon episode has a political, and almost didactic message, it can nonetheless be given to different readings, each canceling the others out while at the same time supporting them. In this manner Arafat-as-symbol was both caricatured and restored.

The same duality that simultaneously ridicules and preserves Palestinian symbols can be traced in Rashid Masharawi's *Upside Down*.[81] The film opens with an old woman, wearing the traditional Palestinian village dress, as she climbs the stairs into an old ruin of a stone house. She enters an old room, whose walls bear faded remains of an indistinct colour, typical of rooms in the old houses abandoned by the 1948 refugees that fill the new Palestinian cinema as recollections of the distant past. The woman prepares the traditional Arab dish called *maqlouba*, with the camera following closely on every detail of the food and the apartment – rice, chicken, lemon, kitchen utensils, stones, walls. This is replaced occasionally by views seen through the window. At first this seems to be an inventory of shopworn Palestinian symbols: village woman, stone house, traditional food, cleaning of rice, dreams of open spaces, floating vision of the al-Aqsa mosque – all representative of the destroyed Palestinian past and the hope for its future revival. However, all of these elements are in the wrong place or are seen from an unlikely angle, much like an unfinished picture puzzle whose pieces are scattered around every which way: the *maqlouba* is seen upside down, through the transparent bottom of the pot; the views seen from the window change continually – mountains, fields, refugee camp, city (perhaps Jerusalem?), the sea. The film nears its end when the *maqlouba* is ready, with the camera showing it from below, and it is then up-turned onto the serving platter – so that it is now right side up. But the serving platter, instead of being placed on a table, begins to soar above Jerusalem and above the mosque, with the empty pot floating above it. The *maqlouba*, like pictures of Arafat, is a national symbol representing and restoring the national past. But when it is floating in the sky it looks ridiculous, like Arafat's picture on the balloon in Suleiman's *Divine Intervention*, out of time and place, and loses its symbolic meaning. However, at the same time, through the effect of estrangement it regains that meaning in a newly vitalised way. Thus, while the film exposes the banalisation of the Palestinian national symbols, it also renews them by giving them a visionary and imaginative surrealistic dimension.

Nizar Hassan's film *The Invasion* can serve as an example of the difficulty in describing the small details of Palestinian life and space without giving them symbolic meanings, as well as an example of possible ways of overcoming this difficulty. The scenes of the destruction in a refugee camp, in Jenin, after its invasion by the Israeli army are cross-edited with what have become much overworked images of destruction in the Palestinian cinema and television networks: a red rose on a tombstone, a close-up shot of a fig leaf in front of a destroyed house, children as emblems of hope, the fighter with the gun, the dead *shaheed*, and so on. However, these tropes are embedded in the cinematic structures and contexts that concentrate on fragments of life, not necessarily on symbols. For example, in an interview with a woman whose fiancé was killed in the raid, the director ignores the symbols of death and the destruction of hope and insists on hearing the small, meaningless details of daily life of the couple: the colour of the settee or the floor plan of their future apartment. Even when the camera pauses by sights and details that have become commonplace, such as the houses destroyed in the raid, it does so in a unique manner. They are first shown as televised images being viewed by the director and the Israeli bulldozer driver he is interviewing. In this way a certain distance is created between the very well-known sights of destruction, which have become symbols for Palestinian fate and those viewing them: the Palestinian director and the Israeli driver. This distance strips the visions of their well known symbolic values. The same scheme is further heightened in the film by its 'detective structure' – only at the end of a sequence or chain of events are certain facts provided – the facts without which it cannot be understood: that the woman's fiancé has been killed, that the ditches viewed by the bulldozer driver are graves, that the child innocently strolling down the camp's street is an orphan. These details, when detached from the entire narrative, cease (for the time being at least) to represent it; they represent reality alone.

Palestinian cinema of the last couple of years bears witness to the continual shrinking of Palestinian space, and has documented the struggle to recover it. This space is by definition an imagined space – indivisible, harmonious and perfect, self-confident and homogeneous. In an interview with Anne Bourlond,[82] Elia Suleiman defined the relations between his cinema and Palestinian nationalism:

> I don't want to relate the story of Palestine, I want to open the way to a multiplicity of spaces that allows for different readings ... I try to construct an image which lies far beyond any ideological definition of precisely what it means to be a Palestinian, one that is far from all stereotypes ... Cinema was not designed to strengthen the national image or its opposite: the negative image of the other.

This multiplicity of dissimilar spaces has become ever more difficult to achieve in the last few years. The *intifada*, as Samir Farid[83] claimed, does not distinguish between men and women, children and adults, rich and poor, but only between Jews and Arabs. Palestinian cinema of this era largely adopts this stance. It is a cinema of exile, but its struggle to appropriate the national territory that is under threat enables only some of its directors of the previous generation, to posit themselves between languages, identities, nationalities and cultures. To a great extent, Palestinian cinema is in search of a lost unity rather than of an open heterogeneity, of one harmonious national space rather than a third multicultural space.

NOTES

1. Safa'hat Min Mothakkaratin Khasba, 1980.
2. See Rashid Khalidi, *Palestinian Identity* (New York: Columbia University Press 1997); Salim Tamari, 'The Local and the National in Palestinian Identity', in Kamel Abdel Malek and David Jacobson (eds), *Israeli and Palestinian Identities in History and Literature* (New York: St Martin's Press 1999) pp.3–9; and Baruch Kimmerling and Joel Migdal, *Palestinians: The Making of a People* (Cambridge, MA: Harvard University Press 1993).
3. Ibid.
4. For this topic, see Edward Said, 'Afterword: The Consequences of 1948', in Eugene L. Rogan and Avi Shlaim (eds), *The War for Palestine: Rewriting the History of 1948* (Cambridge: Cambridge University Press 2002 [2001]) pp.206–19.
5. Issam Nassar, 'Reflections on Writing the History of Palestinian Identity', *Palestine-Israel Journal*, 8/4 & 9/1 (2002) pp.24–37.
6. See also Tamari (note 2).
7. Ibid.
8. Alexander B. Murphy, 'National Claims to Territory in the Modern State System: Geographical Considerations', *Geopolitics*, 7/2 (Autumn 2002) p.193.

9. Edward Said, 'Reflections on Exile', in R. Ferguson (ed.), *Out There* (New York: The New York Museum of Contemporary Art 1990) p.359.

10. Ibid. p.360.

11. Eric Hobsbawm and Terence Ranger (eds), *The Invention of Tradition* (Cambridge: Cambridge University Press 1983).

12. See Edward Said, 'Invention, Memory and Place', *Critical Inquiry* 26/2 (Winter 2000) pp.175–92, who describes similar processes of creating a collective memory at the expense of small units such as the family and the village, that have crumbled. He does not refer directly to the Palestinian society but it is possible that his assumptions are true to it, and relate to the phenomenon described by many, among them, Khalidi (note 2).

13. Said, 'Reflections on Exile' (note 9) pp.357–67.

14. Hobsbawm (note 11).

15. Oren Yiftachel, 'Territory as the Kernel of the Nation: Space, Time and Nationalism in Israel/ Palestine', *Geopolitics*, 7/2 (Autumn 2002) pp.215–48.

16. Ibid. p.215.

17. Ibid. p.216.

18. Like the land itself in the Palestinian culture, the home has become the answer to uprooting and exile. It has become a utopian symbol of the homeland, untouched by the ravages of time and an emblem of national unity, of the land and of the family, see Said, 'Reflections on Exile' (note 9) p.359. On the importance of the home as a private space see David Morley, 'Bounded Realms: Household Family, Community, and Nation', in Hamid Naficy (ed.), *Home, Exile, Homeland: Film, Media, and the Politics of Place* (New York: Routledge 1999) pp.18–41.

19. For the term *third space*, as an intersection between cultures, identities, histories, see Edward Soja, *Thirdspace* (Oxford: Blackwell 1996); Homi K. Bhabha, 'Dissemination: Time, Narrative and the Margins of the Modern Nation' in Homi K. Bhabha (ed.), *Nation and Narration* (London and New York: Routledge 1990) pp.291–323; Hamid Naficy, *An Accented Cinema* (Princeton, NJ: Princeton University Press 2001).

20. Ella Shohat and Robert Stam, *Unthinking Eurocentrism: Multiculturalism and the Media* (New York and London: Routledge 1994).

21. See Gloria Anzaldua, *Borderlands/La Frontera: The New Mestiza* (San Francisco, CA: Aunt Lute Books 1987), who speaks of the possibility of a 'shifting and multiple identity and integrity'.

22. Director of *The Milky Way* (Darb-Et-Tabbanat, 1997) and *The Ninth Month* (Fish-Shahr-Et-Tassee', 2002).

23. Director of *Austura* (Ostura, 1998) and *Independence* (Istiklal, 1994).

24. Director of *The Olive Harvest* (Mawsim Azaytoon , 2003).

25. Gearoid O Tuathail and Simon Dalby (eds), *Rethinking Geopolitics* (New York: Routledge 1998).

26. See also Anzaldua (note 21).

27. Tuathail and Dalby (note 25) p.1.

28. Irit Rogoff, *Terra Infirma: Geography's Visual Culture* (New York: Routledge 2000).

29. Ibid.

30. Tuathail and Dalby (note 25).

31. Ibid. p.3.

32. Ibid.

33. Joanne P. Sharp, 'Reel Geographies of the New World Order', in Tuathail and Dalby (note 25) p.153.

34. For this topic, see Murphy (note 8) pp.193–214.

35. 'Ors Rana, 2002.

36. For the role of journeys in cinema, see Naficy (note 19).

37. 2003.

38. An Ta'bor Qualandia, 2002.

39. Tathkara Ila-l-Quds, 2002.

40. When inhabited by Jews and Arabs living in the same building.

41. Fintithar Salah-Eddin, 2001.

42. As in Elia Suleiman's *Divine Intervention* (Yad Elahiyya, 2002) and in Rashid Masharawi's *Curfew* (Hatta Ish'aarin Aakhar, 1993) and *Haifa* ('Haifa, 1995).

43. Tawattor, 1998.

44. Among them the previous films of Masharawi himself, *Haifa* and *Curfew*.

45. Thalathato Centimetratin Akall, 2002.

46. Ijtiah, 2003.

47. Waka'ee Sijill Ikhtifaa', 1996.

48. About the way to create a connection, a sense of belonging or ownership of the landscape, see W.J.T. Mitchell, 'Imperial Landscape', in W.J.T. Mitchell (ed.), *Landscape and Power* (Chicago, IL and London: The University of Chicago 1994) pp. 5–34.

49. In a private interview in Ramallah, 2002.

50. In a private interview in Jerusalem, 2003.

51. Sharp (note 33) p.153.

52. Hodoudo-l-Holm Wal-Khaouf, 2001.

53. Anton Shamas, 'Autocartography: The Case of Palestine, Michigan', *Palestine–Israel Journal* 9/2 (2002) p.114.

54. For this subject, see Shohat and Stam (note 20). The idea of re-territorialising the landscape is based on Gilles Deleuze and Felix Guattari, 'What is Minor Literature', in Russell (note 9) pp.59–71. See also Rogoff (note 28).

55. Naficy (note 19).

56. Frederic Jameson, 'Third World Literature in the Era of Multinational Capitalism', *Social Text* 15 (Fall 1986) pp.65–88.

57. This term will serve us freely. In contrast to his declarations, defining the allegory of the Third World as a polysemic allegory, heterogeneous, multidimensional, ever changing, in Jameson's specific analyses, he reveals a one-dimensional relationship between the literary or cinematic signifier and the meaning it represents, which for him, in the third world, is always a national significance, see also Aijaz Ahmad, *In Theory* (London and New York: Verso 1992). We will distinguish here between such a one-dimensional allegory, in which the personal experience stands for the nation and a multidimensional allegory, which constitutes an intersection of national, class, gender, universal meaning, which add on to concrete details, without replacing them. The term symbol will serve us in the same way. We will apply it to objects in reality which represent, in a one-dimensional or a complex way, the national existence, or a heterogeneous variety of additional existents. For the subject of symbol and allegory, see Shohat and Stam (note 20).

58. W.J.T. Mitchell, 'Holy Landscape: Israel, Palestine, and the American Wilderness', *Critical Inquiry* 26/2 (Winter 2000) pp.193–224.

59. As well as in those of other directors.

60. See for example Muhamad Bacri's *1948* (1998); Aza El Hassan's *Place/ Out of Eden* (Makan/ Aw Kharij-1-Janna, 2000); Liana Bader's *Fadwa* (1999), *Zaytounat* (2002) and *Green Bird* (At tayr-Ol-Akhdar 2002); Azza El Hasan's *Three Centimeters Less*; Nizar Hassan's The Invasion; Elia Suleiman *Divine Intervention*.

61. For example in Rashid Masharawi's *A Ticket to Jerusalem*.

62. Naficy (note 19).

63. Though the mosque is a religious symbol it has been appropriated by secular filmmakers, who have adopted the religious stance increasingly prevalent in Palestinian society.

64. Barbara Mckean Parmenter, *Giving Voice to Stones: Place and Identity in Palestinian Literature* (Austin, TX: University of Texas Press 1994) p.74.

65. The cinematic way to turn the olive tree into a fetish is by freezing its frames, by repetitions of the same shot, by taking it out of context or by slow motion photography.

66. In Liana Bader's *Fadwa*, and *The Green Bird* and inSubhi A-Zubeidi's *Crossing Qualandia*.

67. Hikayatil Jawaher Eth Thalath, 1994.

68. Michel Khleifi, 'Orso l-Jaleel, 1987.

69. Subhi A-Zubeidi, Ad Daw' Fi Nihayatin-Nafak, 2002.

70. See note 39.

71. See note 52.

72. See note 60.

73. Raja Shehada, *When the Bulbul Stopped Singing: A Diary of Ramallah Under Siege* (Profile Books 2003) p.86.

74. Edward Said and Jean Mohr, *After the Last Sky: Palestinian Lives* (New York: Columbia University Press 1999 [1986]) p.4.

75. Elias Sanbar, 'De L'identite culturelle des Palestiniens', in Elias Sanbar, Subhi Hadidi and Jean Claude Pons (eds), *Palestine: L'enjeu culturel* (Paris: Circe, institute du monde arabe 1997) pp.26–7. See also Said, 'Afterword' (note 4).

76. Michel Khleifi, 'La force du faible' (Entretien avec Michal Khleifi), *Cahier du cinema* 87/3 (November 1987).

77. Al Holmo-l-'Arabi, 1998.

78. Jean Baudriard, *Simulacres et Simulation* (Paris: Galilee 1981).

79. Frederic Jameson, *Postmodernism or The Cultural Logic of Late Capitalism* (Durham, NC: Duke University Press 1991).

80. On this matter, see Naficy (note 19).

81. Maqlouba, 2002. Maqlouba, Arabic for upside down, is the name of a traditional national dish which, when ready, is upturned from the pot onto the serving dish.

82. Anne Bourlond, 'A Cinema of Nowhere: Interview with Elia Suleiman', *Journal of Palestine Studies* 29/2 (Winter 2000) pp.95–101.

83. Samir Farid, *The Palestinian Cinema and the Conquered Land* (Cairo: Collection of Cinema 1997) (in Arabic).

Inscribing the American Body Politic: Martin Sheen and Two American Decades

SIMON PHILPOTT
School of Geography, Politics and Sociology, University of Newcastle, UK

DAVID MUTIMER
Centre of International Cooperation and Security, University of Bradford; Department of Political Science, York University, Canada

It is hard not to think about the military flavour of American political desire when you sit, as we did one balmy Toronto day in August 2001, engrossed for more than three hours in the remarkable *Apocalypse Now Redux* (*ANR*). The film centres on the bravura performance of Martin Sheen as Captain Benjamin Willard, but one finds oneself seeing in the young Sheen Jeb Bartlet, US president in the popular political drama, *The West Wing*. As we reflected over a beer in a downtown pub on the film we had just seen, we began to wonder about the kind of America for which Sheen's performances provide bookends. *ANR* and *The West Wing* mark out two crucial moments in the American imperium. Vietnam is synonymous with the decline of American power, seemingly the point at which the United States not only exceeded the limits of its reach but also degraded itself in places like My Lai and Cambodia. *The West Wing*, by contrast, is a vision of America resurgent in which the post-Cold War United States bestrides the globe and, significantly, is recommitted to a liberal vision of the American Dream as it looks toward the twenty-first century. Two terms of the Clinton presidency of domestic prosperity and cautious commitments to multilateralism stood between the Gulf War and that August afternoon and seemed to herald a United States more at ease with itself and less restless in international affairs. Moreover, there seemed every reason to be optimistic that the emerging but bumbling Bush presidency of classical US isolationism and unilateralism would endure for just one term. In short, despite the policy and moral failings of the Clinton presidency, it seemed plausible to imagine that American voters were moving in a direction of reconciling themselves to an America of less grandiose imperial vision than had characterised the Reagan and Bush senior years. We had no idea what lay just around the corner.

Now the United States is in a more bellicose mood than perhaps ever in its history. After September 2001 Republican unilateralism took on an all-encompassing aspect with Bush's neat division of the world between those who would follow an American lead and those who would place themselves against the United States. Rather than emphasise its commitment to the path of grey, patient, forensic police work to establish responsibility for the 11 September (9/11) attacks and to interrupt the flows of finance, money, people and materiel that sustain terror networks, the Bush administration, unsurprisingly, declared war on terror and threatened it upon states it deemed in any way supportive of terror. The inevitable and early invocation of discourses of identity, self-sacrifice, and moral righteousness seemed instantly hyper-real yet far from frivolous. Ominously for those perceived as enemies by the United States, its vengefulness was characterised by an unsubtle mix of innocence wronged and a gunfighter nation seeking dead or alive justice.[1] Arguably, the attacks on New York and Washington represent significant steps in the re-imagining of innocence lost in revolution, civil war, slavery, murderous campaigns against indigenous peoples, and in Vietnam. While innocence lost cannot be reclaimed, nationalist discourse often attempts to do just this by effacing the violence of nation-making and emphasising glorious sentiments and deeds in the Sisyphean project of establishing the uniqueness and (moral) purity of nations.

It is the relationship between moral innocence and the capacity for self-sacrifice in its protection that is of interest to us in our analysis of the Americas depicted by Martin Sheen in that brief moment when his two most significant roles, the assassin, Captain Benjamin Willard, in *ANR* and President Josiah Bartlet in *The West Wing*, existed simultaneously in everyday popular cultural awareness. The complexities of this relationship should not be underestimated. Citing Gore Vidal, Mark Lacy argues that the twin desires of the United States to dominate global affairs and yet to be seen as completely innocent leads to distortions of history and the forgetting of its lessons.[2] Moreover, it is a matter of some irony that the issue of sacrifice has become so central to the US Presidential campaign underway as we conclude final revisions on this article. That is, from the vantage point of September 2004, it appears that Democratic candidate John Kerry's Vietnam War record has successfully been used to attack his credibility by an incumbent President who, it is clear, used family connections to avoid military service of almost any kind, let alone a tour of duty in Vietnam. To this end, George W. Bush's arriving on the US aircraft carrier USS *Abraham Lincoln* (arguably a president with a significant understanding of sacrifice in defence of principle) by fighter plane and dressed as an airman to announce the end of major combat operations in Iraq in May 2003 suggests a simulacrum of self-sacrifice. It is also indicative of the way in which sacrifice remains contested in contemporary American electoral discourses and points to the relevance of analysing its representation in popular culture. As Lacy argues, cinema and, we propose, popular culture more generally, is important in the active forgetting and active production of history.[3] Moreover, representations of the two 'crews' that surround Sheen in *ANR* and *The West Wing* tells us much about issues of gender, race and class in American politics. On the surface there seems little to connect the men who accompany Willard up river in *ANR* with those who support Bartlet's journey through the American presidency in *The West Wing*. However, despite the obvious differences between the seemingly impoverished grunts on Willard's mission and the powerful, educated and wealthy men of Bartlet's inner circle in *The West Wing*, there are enduring themes in the ways that gender, class and race are represented.

In providing this reading, we seek to add to the growing scholarship on popular culture and (international) politics.[4] While neither would seem to be a particularly convincing representation of their particular realities, *ANR* and *The West Wing* have, respectively, profoundly influenced popular understanding of the Vietnam War[5] and the workings of the presidency. For example, with a viewing audience of more than 18 million per week, *The West Wing* reaches more Americans than any regularly scheduled news broadcast.[6] On the one hand, as former Clinton press secretary and consultant to *The West Wing* Dee Dee Myers argues: 'Pop culture is a pretty good barometer of what resonates with the people … When you tune into Jay Leno and hear a joke about Clinton getting a haircut on the tarmac, you know that isn't good.'[7] But on the other hand, the popularity of *The West Wing* is in part based on admiration for the saccharine morality of the Democrat Bartlet and a sentimental desire to negate the odious aspects of the Clinton presidency.[8] Moreover, there has long been an understanding between Hollywood and the US government (and its various agencies) about Hollywood's capacity to manufacture support for US war efforts. Indeed, in the wake of 11 September, George Bush's senior aide, Karl Rove, met with Hollywood's top brass to enlist its help in prosecuting the war in Afghanistan.[9] In a more general sense, the electronic media, especially television, introduces another layer of experience for the subject that complicates the possibility of 'truth'. Electronic mediation and repetition of particular images destroy the relationship between sign and referent leading to a condition of hyper-reality in which images take on a distinct existence and meaning.[10] On this view, *ANR* and *The West Wing* provide examples of the way that mediated communication draws upon and transcends particular realities.[11] For example, members of Congress test policy ideas through *The West Wing's* scriptwriters, guaranteeing both their airing and 'intelligent mulling … more perhaps than [they] might be assured in the real world'.[12] Therefore, the underlying assumption of this article is that popular culture matters and provides the basis of what many people believe about the world in which they live.

'THE HORROR, THE HORROR...'

Conventional wisdom has it that the war in Vietnam marked an end to American innocence and forever changed US attitudes to international conflict. While this simple approach overlooks the possibility that the loss of innocence is a recurrent feature of US politics and identity discourse, it is a useful reminder that forgetting or externalising responsibility for past violence is integral to the making of national identity.[13] In other words, innocence can only be lost once but the active forgetting of its loss creates the conditions for its being lost over and again. This repetitious loss of innocence has left its marks on the US polity. For example, some argue that one consequence of the so-called Vietnam syndrome is reluctance on the part of the US to use military force.[14] However, even prior to large-scale offensives against Afghanistan and Iraq, the United States proved itself willing and able to perpetrate violence against other countries. The syndrome is perhaps more accurately associated with an unwillingness to suffer casualties – violence is fine, but not American body bags. The supposed corollary of the syndrome is rapid technological development in military affairs enabling the United States to wage war with minimal casualties, although films as different as *First Blood* and *Black Hawk Down* seem to suggest determined human beings can triumph over technology.[15] A willingness to deploy military force necessarily implies at least the possibility of casualties, even in the asymmetrical conflicts that characterise recent American military operations, so consideration of the limits of the United States' incapacity to act and what it is and is not willing and able to sacrifice are important.

It seems possible to locate the precise moment of the death of American innocence and the birth of the Vietnam Syndrome in *ANR*. Colonel Walter E. Kurtz (Marlon Brando), deranged and waging a private war using Cambodian-based Montagnard tribesmen, recites the eulogy for the young United States towards the end of the film, explaining to Willard why it cannot win the war.

> We went into a camp to inoculate the children. We left the camp after we had inoculated the children for polio, and this old man came running after us and he was crying ... We went back there and they had come and hacked off every inoculated arm. There they were in a pile ... a pile of little arms. And I remember I cried. I wept like some grandmother. I wanted to tear my teeth out. I didn't know what I wanted to do. And I want to remember it. I never want to forget it. I never want to forget. And then I realised ... like I was shot. Like I was shot with a diamond. A diamond bullet right through my forehead. And I thought my God, the genius of that. The genius. The will to do that. Perfect, genuine, complete, crystalline, pure. And then I realised they were stronger than we. These men who fought with their hearts, who had families, who had children, who were filled with love, but they had the strength

> to do that. If I had 10 divisions of those men our troubles here would be over very quickly.

In Kurtz's view, the purity of intention animating violence is all-important. The United States' problems stem not from an unwillingness to employ violence, rather, the United States has lost its moral compass, rendering its violence ineffective and a measure of its lassitude and corruption: it is unwilling and unable to make the kinds of sacrifices that are necessary to succeed.

We suggest that Kurtz and Willard in *ANR* provide an important gloss on the Vietnam syndrome. The issue is not that the United States is unable to kill, main or otherwise perpetrate violence. It is clear that the United States, in Vietnam and afterwards, was more than willing and able to perpetrate violence. Rather, it is the sacrifice of the *self*, particularly the *moral* self, that is missing in the United States. Kurtz recognises the need to act in ways that the US military does not and cannot permit; he takes it upon himself to do what his country cannot do, and it costs him himself. Willard, sent by the US military to kill Kurtz, comes not only to understand Kurtz's choice and its costs, but also to be capable of the same sacrifice. His voyage up the Nung River leads to his own moment of choice, in which he must sacrifice himself to kill Kurtz. Willard recognises that he cannot murder Kurtz and remain within the US military, even though it was that military that sent him to do the job: 'They are going to make me a full major for this, and I'm not even in their fucking army anymore'. Willard's removal of himself from the army is more than simply a decision to leave; it is the sacrifice of himself just as it was before him for Kurtz. At the very beginning of the film, Willard exposes how completely his self is merged with the military: 'I took the mission … what the hell else was I going to do?' The moment of sacrifice is, therefore, a moment poignant with Willard's own self-awareness and his recognition of the (institutional) self-delusion of the US Army, earlier confirmed by a young Colonel (Harrison Ford) informing Willard that the mission to assassinate Kurtz does not exist nor will it ever exist.

Kurtz and Willard mark the boundary of the (im)possible for the United States because they are willing to sacrifice their moral selves to what they think/know needs to be done. The self-sacrifice of Kurtz and Willard occur outside the United States because the latter is unable to confront itself in acts of murder, which are, of course, inconsistent with the self-delusion that the United States is an indivisible and honourable force for good. The Vietnam syndrome is now supposedly expunged,[16] but if the syndrome rests in the incapacity to sacrifice the moral self to what is necessary for action, it is an open question whether it has been overcome. Certainly, the United States is willing to act internationally, but as has been remarked, not at the cost of large numbers of American lives. Does the aversion to self-sacrifice run deeper than this, however?

Seen as a profound hesitation in the capacity to lose the moral self of the United States, the Vietnam syndrome has recently played itself out in a very public discussion of the legacy of the war. In April 2001, the *New York Times* reported that Senator Bob Kerrey had committed war crimes while in charge of a small group of Navy Seals in Vietnam.[17] While he denied his culpability for war crimes, he did admit to killing civilians with the consequence that, 'in the senator's own words: "I went out on a mission and, after it was over, I was so ashamed I wanted to die."'[18] Not surprisingly, when this story broke the news media turned, among others, to Kerrey's colleague in the US Senate, Senator John McCain, a man who had spent years in a Vietnamese POW camp, and who had, the year before, run unsuccessfully for the presidency. In his defence of Kerrey, who he argued was a war hero, not a criminal, McCain made it clear that he, at least, had come through his own voyage up a Vietnamese river to a hellish camp with his capacity for the sacrifice of the moral self very much intact:

> McCain argues that Vietnam was the kind of war that required its participants to hate the enemy, and he unabashedly makes a combat virtue out of hate. In his words: 'I hated my enemies even before they held me captive because hate sustained me in my devotion to their complete destruction and helped me overcome the virtuous human impulse to recoil in disgust from what had to be done by my hand.[19]

For his part, Martin Sheen, in his incarnation as President Jeb Bartlet, must also grapple with the relationship between identity and action and cannot step outside 'this man's army' as Willard did. The question of moral sacrifice is confronted directly in an episode entitled 'A Proportional Response' from early in the first season (aired 6 October 1999). In this episode, Bartlet must make his first decision as commander-in-chief, and in doing so confronts directly the limits of the possible in the capacity of the United States for self-sacrifice. When an American aeroplane is shot down in the Middle East and American lives are lost, it is clear to the president and his staff that the United States must respond. The military sets out the options for the president, a series of 'proportional responses' of the kind with which we are familiar: taking out an airport or a military base. Bartlet rages that it is not sufficient and that for the crime of killing Americans (including his personal doctor who had a young wife and child), the full weight of the United States' tremendous power should be visited on the perpetrators. What was needed, he exclaims, is a *disproportionate* response: the use of extravagant violence to make a point.

At Bartlet's direction, the Pentagon draws up a list of possible targets whose destruction would inflict far higher costs. The problem, of course, is that such a large-scale response would cause massive 'collateral damage', hundreds of civilian deaths. Here then is the moment at which the contemporary United States must face its capacity to sacrifice its moral self, to will itself to the kind of murderous actions undertaken by Kurtz and Willard in *ANR*. However, when faced with the need to make this sacrifice of the moral self of the United States, Bartlet hesitates and ultimately refuses. He searches for justification, asking why it is that the United States makes a proportionate response, must always make a proportionate response. It is Leo McGarry (John Spencer), his chief of staff, who answers him: 'because that is what we do'. He could as easily have said: because that is who we are. The identity of the United States is intimately tied with these kinds of responses, imposing limits on the possibilities of action because of who the United States is. In *The West Wing*, therefore, Sheen's actions are again influenced by the limits of the American (im)possible.

The 'we' of *The West Wing* remains the 'we' of the self-limited United States unable to do what was necessary to win the war in Vietnam. Nonetheless, the limits of this moral self still allow for terrifying violence directed against others. The Pentagon and the White House decide what violence is permissible – how much violence is 'proportionate' – and in this respect the United States' understanding of itself as an innocent power is delusional. Thankfully, the United States did not bring the full weight of its power to bear in Vietnam, or more recently in Afghanistan, given that the nuclear option was canvassed by more extreme elements among the US citizenry in both cases. The severity of even a limited response, however, for those on the receiving end was dramatically demonstrated in 2003's invasion of Iraq. Here a US force, deliberately kept smaller than some analysts demanded, but pointedly involving US ground-forces from the outset, wreaked heavy destruction on Iraq and Iraqis.[20]

The loss in Vietnam, which occurred in a way that called into question the capacity for *sacrifice*, also called into question American masculinity, in part because military violence in particular has long been intimately connected with the production of masculinity. For example, Kurtz's transformation from 'a tough motherfucker' (who completed gruelling Green Berets training aged 38) to obesity suggests an intimate link between an America unable to sacrifice its moral self and an America grown flabby and corrupt in the comforts of its continental isolation.[21] Recovering from Vietnam, therefore, has entailed a reconstitution of American masculinity, much of which has occurred in popular culture.[22] Arguably, the murder of *ANR's* Kurtz contributes to the end of the angst-ridden masculinity of the counter-culture era and foreshadows the reinvention of Hollywood-defined masculinity in the Reagan era. The contrast between Brando's Kurtz and another Vietnam veteran turned 'irregular', Sylvester Stallone's John Rambo, could not be starker.[23] To explore questions of American masculinity, we now turn to a more detailed examination of the people around Sheen on the Nung River and in *The West Wing*.

MEN, BLACKS AND SERVANTS IN AMERICA

Linda Boose[24] observes that American men of the Vietnam generation squandered an opportunity to define their masculinity away from the battlefield. Had they done so, they would have been the first generation of twentieth-century men who had proven themselves around the home, rather than far from it. Many followed willingly in the footsteps of their grandfathers, fathers and uncles who had served in the American Civil War, two world wars and Korea. But others were conscripted to fight in Vietnam, greatly sharpening American public unease about the justice and morality of the war in Vietnam. Moreover, the war in Vietnam raised questions about American masculinity, in part reflected in films like *ANR* and *Platoon* that showed the destructive effects of extraordinary violence on ordinary men. Whether remembered as wimps or baby-killers, Vietnam veterans seen as having brought shame upon themselves and their nation, were treated scornfully by supporters and opponents of the war alike, although narratives of vilification underwent revision in the wake of the 1991 Gulf War.[25]

In the Reagan era, Hollywood's struggle to reconstitute masculinity had particular effects in cinematic representations of the male body with a simple, muscular, no-nonsense, white masculinity increasingly common.[26] Through Arnold Schwarzenegger, Sylvester Stallone and Bruce Willis, among others, heroic figures were created to reaffirm a traditional if infantile version of masculinity and to liberate it from Vietnam. These muscular 'hard bodies' resort to physical aggression and intimidation in their engagement with others and problems in their cinematic worlds.[27] One, Stallone, even returned to Vietnam to rewrite the ending: 'Do we get to win it this time?'[28] Another, Schwarzenegger, has taken at least the language of his 'tough guy' persona from screen to electoral politics, not hesitating to imply that the unflinching characteristics with which he is associated animate his search for solutions to political problems and impasses.[29] For now, he enjoys significant support among sections of the Republican Party more for what he represents than what he has thus far achieved. The new hard men literally and figuratively filled the screens of the 1980s, the number of roles for women falling sharply.[30] Interestingly, films that construed themselves as anti-war tended to be peopled with smaller men such as Michael J Fox, Tom Cruise, Chris Sheehan and Willem Dafoe.[31] If the Vietnam generation asked provocative and unanswerable questions about modern American masculinity, it was Stallone and Schwarzenegger, corporeal metaphors for a beefed-up American foreign policy promoted by Reagan and the Republican right, who provided the answers. Indeed, when uniformed veterans protested outside a Boston opening of *Rambo* in the mid-1980s: 'According to the teenagers screaming at the vets to go home, it was Sylvester Stallone – a real man – who was also a "real veteran."'[32]

The crew of the small river patrol boat in *ANR*, Chef Hicks (Frederic Forest), Lance B. Johnson (Sam Bottoms), both white, Mr Clean (Laurence Fishburne) and Chief Phillips (Albert Hall), both black, are the kind of men and veterans that so upset the Boston teenagers. In individual ways they are cast as victims of the US military, rather than the conflict itself. All of them would willingly go home, but, as Willard observes: 'Trouble is, I've been back there, and I knew it just didn't exist anymore.' This suggests the psychological dissonance each of the crew experience is fed by their at least partial awareness that the United States is in the midst of a profound identity crisis and unsure about its feelings and obligations to its soldiers. None of the crew, with the possible exception of Chief, the boat's captain, is overtly muscular. Rather each has the lean body of youth, albeit worked over by the US military's training programmes. Of Clean, Chef and Lance, none has interest in, or aptitude for, the conflict in which they find themselves. Only Lance, a champion surfer from southern California, and Sheen's Willard, survive the arduous journey up river. But Lance's survival is entirely fortuitous. Moreover, Lance's disengaged recklessness contributes to the deaths of Clean and Chief. Lance smokes dope, skis behind the boat as it travels up river, murders Vietnamese civilians aboard a river boat in a moment of panic (also involving Clean), takes LSD and, as the film reaches its murderous climax, is seen to be on the brink of 'going native' and losing touch with his militarised 'self'. Lance could easily have met the random deaths of Clean and Chief, but his ability to avoid lethal fire might be Coppola's way of reminding the audience about the disproportionately high numbers of blacks who were at the front lines of conflict in Vietnam.[33] Arguably, Lance's narcissistic selfishness is presented as evidence of the moral failings of 1960s youth culture, of which, as a surfer, he is part. The absence of conscience – his main attachment throughout the film is to a puppy 'orphaned' by his murder of its Vietnamese owner – suggests alienation, not only from his duties as a soldier, but also himself.

Paradoxically, Lance is valorised by Lt Col. Bill Kilgore (Robert Duvall) whose uncompromising prosecution of the war with Air Mobile is intertwined with his endless search for good surfing beaches. Kilgore, seemingly respected by his men, displays no obvious reverence for the institutions of the military, ignoring his responsibility to cater to the needs of Willard's mission in favour of discussing surfing with Lance. Kilgore's matter-of-fact, unyielding toughness is in stark contrast to Lance's wide-eyed naiveté and self-preservation instincts, yet both are curiously detached from a broader view of the war, the military itself and US foreign policy goals. Thus the 'authority figure' Kilgore, who routinely uses racist epithets and seems addicted to the pleasures of war, the second most senior officer we encounter in the field after the deranged Col. Kurtz, gives succour to the ill-disciplined narcissist Lance. Willard's ambivalent admiration for both Kilgore and Kurtz highlights his recognition of the unconventional means necessary to prosecute the conflict successfully and his detestation of the ersatz morality of the military hierarchy in whose employ he is. All three clearly understand that the United States' struggle with the idea of its continuing innocence is delusional. Yet all three have created individual moral realms that sufficiently distance themselves from those that they must kill (and have killed) and that enable them to do so.[34]

Chef is different to Lance and Kilgore, 'wrapped too tightly for Vietnam', observes Willard; 'probably wrapped too tightly for New Orleans', he adds (New Orleans being Chef's home). Of the five men making their way into the Cambodian heart of darkness, it is Chef who exists on the brink of psychological disintegration. His fear and loathing are rarely far below the surface of his brittle exterior and we learn of his training as a saucier: both occupation and character serve to feminise Chef. In a moment of temporary unravelling Chef shrieks 'I want no part of it, you can kiss my ass', indicating his awareness of the moral abyss the war has become. Yet for his psychological stress and palpable fear, it is Chef that is the best of the crew. Unlike Lance and Clean, Chef does not engage in undisciplined acts of violence against Vietnamese civilians. Perhaps ironically, Chef dies at the behest of Kurtz, both of whom are gravely affected by the pressures and moral contradictions of the war in Vietnam. In having him murdered, Kurtz takes the life of the one-crew member who has not directly contributed to 'the horror', as Kurtz describes it at his own moment of death.

It is tempting to link the survival of Lance and Kilgore to their amorality and ability to do what it takes (to survive). Neither is visibly affected by the suffering of the Vietnamese nor concerned by the extraordinary, technologically sophisticated violence that they use. Indeed, Kilgore's lament that 'some day this war's gonna end' indicates he is singularly untroubled by moral issues. Kurtz and Chef, however, not only see the ineptitude and hypocrisy of the US war effort, but also personally suffer the consequences of the United States' moral agonising and the disintegration of its political will. Willard, Lance and Kilgore, alienated and seemingly motivated by self-interest, survive. Their individualism appears at odds with what is required for the US military to succeed in Vietnam and yet is oddly consistent with the broader ethos of American capitalism. In transcending the political failings of US policy in Vietnam and 'doing what is necessary', they embody its immorality.

It is Chef and Lance that Schwarzenegger and Stallone are produced to overcome. These are men of firm, directed, unquestioning action, certain of their rightness and wrapped in the US flag of Reagan-era muscularity in foreign policy. Yet, by the late 1990s, their time had passed. Their successors, Josh Lyman (Bradley Whitford) and Sam Seaborn (Rob Lowe), are found in the White House of Jeb Bartlet. These are tertiary-educated men whose masculinity is defined by sharp intelligence – lap-tops, not guns – and whose bodies bear no signs of acquaintance with gymnasia or weight rooms.

They are as self-assured, articulate, and morally and socially aware as their Vietnam forebears Chef and Lance, were not. They supposedly embody the rebirth of a Democratic America and its dreams of justice, equality and fairness. However, in our view, Bartlet's White House suggests that there has been a systematic forgetting in the United States since Sheen travelled up river to murder Kurtz, so that the United States of *The West Wing* is not the mature self, having integrated its loss of innocence, but rather an immature self who, in the decades since Vietnam, expunged its memory to (re)present itself as innocent.

The deaths of Clean and Chief Phillips in *ANR* leaves the story in white hands and blind the surviving whites to the racist logic of Kurtz having a large Montagnard community under his spell.[35] Bartlet's White House is likewise notable for the absence of blacks in the inner circle, except, of course, Bartlet's 'body man', Charlie Young (Dulé Hill), whose role is to serve, to keep control over the domestic business of the White House, but not to think.[36] Charlie is the liberal black ideal type. He is street savvy, understands the sense of belonging that inner city blacks get from gang membership, and is ambitious. The little we know of his family suggests that they are model minority in so far as they have engaged in a good-natured struggle for self-improvement. Charlie is fatherless. We know this because when Charlie's mother, a police officer, suffers a premature violent death 'in the line of duty', Charlie is orphaned. The absence of a father and the violent death of his mother also make Charlie a 'typical' black. Moreover, at the beginning of the series, he is not at college, but working because he is raising his younger sister.

Charlie is *The West Wing's* attempt to show how far race relations have improved in the United States since Clean's own premature and violent death in the line of duty, but it is largely a liberal illusion. One of the reasons Charlie gets the job as Bartlet's body man is because of a sense of liberal *noblesse oblige* among the white staffers (Charlie originally applied for a job as a messenger and is insistent that that is what he is qualified and able to do when pressed by Josh and Sam to take on the job as the President's body man). Charlie fulfils the ethnic fantasies of many liberals: a young, self-possessed, sober, drug-free black man taking charge of his battered family and all the while bettering himself. This is the kind of black man that liberals like Josh and Sam know they can whiten and integrate into their world. As Herman Gray argues: 'African Americans are constructed as invisible and hence ordinary as long as they conform. They become "black" and thus dangerous when they transgress.'[37] On this view, Charlie proves to white Americans that their suspicions about blacks and black families were correct all along. That is, Charlie is living evidence that the experience of disadvantage does not have to determine one's life. Individual will and responsibility for one's self and one's family, are the keys to success and demonstrate that while all may not start off in ideal circumstances, all are afforded equal opportunity if they are willing to grasp it. Thus, *The West Wing* appears to endorse arguments long common in American society. For example:

> In March 1965, Lyndon Johnson's assistant secretary of Labor, Daniel Patrick Moynihan, published a Report on the Black Family which laid much of the blame for black poverty on the 'tangle of pathology' of the black family. He admonished African Americans to rehabilitate their dysfunctional families in order to achieve economic and social assimilation.[38]

Charlie represents the ideal black type envisaged in Moynihan's report. Charlie might look black and occasionally even speak as a black (never like a black), but to all intents and purposes, his presence in the White House serves as constant confirmation to whites that ultimately, there are few insurmountable barriers to success. Charlie's appointment to the inner-circle allows those with real political power to balance their commitments to the ethnic, class and gender disadvantaged with an equal belief in individual self-responsibility. Moreover, Charlie's easy manner with and daily proximity to the president obscure the fact that he is a junior $600 per week employee of the White House.[39]

If Charlie provides an example of a young black man earning a salary below the Washington average,[40] Sam and Josh represent the opportunities of post-Vietnam masculinity for white men. They embody much that is good about America, especially in their discourses of care for the American self (interestingly at odds with Rob Lowe's earlier dissipated lifestyle). In their vitality, articulateness and boyishness they prove what care of the self can deliver, to individuals and to the nation. They see themselves as progressives and locked in perpetual struggle with their Republican nemeses. They profess care for the environment, for education, for the poor, for abortion rights; believe in redistribution of wealth through just taxation, gay rights, gun control, support the elevation of minorities to key posts, and express antipathy to the tobacco industry. Indeed, they provide relief from the disappointments of centre-left politics in real life![41] Their banter is learned, their humour sharp, their manner pompous. Both strut, show off and are profoundly conscious of their power. As representations of liberalism perfected in the United States, they are reasonably open-minded and routinely demonstrate a capacity to learn from errors and to be persuaded by sound arguments. Yet despite their education and rapier-like intelligence, they manifest fairly predictable white, male, middle class, heterosexual prejudices and are often forced to eat various kinds of (minority) humble pie.

Sam and Josh, more than any other members of the cast, are the living embodiment of America's desire to believe in its innocence. While the nation may have sacrificed 58,000 mostly ordinary young men in Vietnam, in Sam and Josh we see the fruits of the nation's care for young men of high aspiration and intelligence. They are the proof that America can do right by its young people and has learned from the needless sacrifice of men like Chef, Clean and Chief Phillips. Their national characters are complex, though. They see themselves and are represented as good men, willing to fight for the things that they hold to be important. The manner in which they achieve their goals seemingly takes us a long way from the White Houses of Nixon and Reagan with their Watergate and Iran-Contra scandals. They also remove us from the unsavoury aspects of the Democratic White House of Bill Clinton.

But, as the collective face of contemporary American masculine identity, Sam and Josh conform to what Lynda Boose calls 'the Boy eternal'. Here is an eternal innocence, a 'boyishness' that cannot take the crucial step to adulthood by turning and examining itself in a critical way. The United States, as represented by two young insiders in the Bartlet administration, confronts eternality in the way Nietzsche warned against:

> What if a demon crept after you one day or night in your loneliest solitude and said to you: 'This life, as you live it now and have lived it, you will have to live again and again, times without number; and there will be nothing new in it … The eternal hour-glass of existence will be turned again and again – and you with it, you dust of dust!' Would you not throw yourself down and gnash your teeth and curse the demon who thus spoke? Or have you experienced a tremendous moment in which you would have answered him: 'You are a god and never did I hear anything more divine!'[42]

Arguably, Sam and Josh embody the blindness of eternal American innocence to issues of race, class, sexuality and gender. Josh greatly benefits from the loving devotion of his aide, Donna Moss (Janel Moloney), but occasionally she is compelled to pull Josh into line for his minor abuses of her. These take the form of sharp talking, forgetfulness, contradictory instructions, intolerance and other office politics misdemeanors.[43] Consider, too, Josh's realisation in negotiations with a gay Republican congressman who supports the Marriage Recognition Act (which will only recognise unions between men and women) that to be gay is not necessarily the defining characteristic of one's life. As the congressman says to Josh, 'My life doesn't have to be about being a homosexual – it doesn't have to be entirely about that.'[44] This is an epiphany for Josh, who suddenly realises the narrowness of his thinking on gay issues. But, of course, being a good liberal, he learns and moves on.

Similarly, Sam's Eliza Doolittle treatment of a call girl, Laurie (Lisa Edelstein), for whom he falls in the first season, betrays a similar insensitivity to otherness. While Laurie represents herself as making difficult but rational choices in her quest to put herself through law school, Sam takes upon itself to try and 'save' his lover from herself but ultimately fails as she refuses his overtures of financial assistance, choosing independence and control over her own destiny. Despite Sam's best attempts to humiliate Laurie publicly, she rebuffs him and delivers him a painful lesson. Moreover, Sam's professional loyalty to the White House Chief of Staff, Leo McGarry, leads him to accept Josh's plan that they put the squeeze on Laurie for information about Republicans with whom she may have had sex.

Laurie is fair game in the mind of Josh and Sam because ultimately her relationship with Sam falls outside of liberal norms and so is stripped of legitimacy and privacy. The inability of Sam and Josh to accord the woman dignity and respect enables them to trample upon Laurie's private life in the quest for political advantage over their Republican opponents. That is, when push comes to political shove, her subject position as a 'hooker' is rapidly exposed and becomes the site of mild zealotry on the part of Sam and Josh. Paradoxically, this occurs in an attempt to protect the dignity and privacy of the recovering alcoholic Leo McGarry, who in turn strongly chastises the two younger men for pressuring Laurie and compels them to return to apologise to her because this is 'not what we do'. Their ready willingness to disregard the public–private distinction, often regarded as central to the liberal social contract, in the face of otherness is indicative of the limits of liberalism's inclusiveness. While Charlie fulfills the ethnic 'model minority' stereotype of Sam and Josh and so 'fits' in Rawlsian terms, Laurie, the 'whore', does not and is pushed to the margins of liberal acceptability.[45]

In neither of two significant scenes involving women in *ANR* are they seen to be in control of their circumstances. In one, Sheen's boat happens across a dysfunctional medivac camp and he learns that two Playboy bunnies that had earlier entertained troops at a surreal and chaotic show at Hua Phat are trapped there. Caught in a particularly unpleasant *cul de sac* of the war, the women are essentially sold into sexual service by their manager in return for the fuel that will enable them to leave the camp. The two women are in poor psychological states, but this does not prevent the crew-members from falling upon them for episodes of quite fetishistic sex. Moreover, both the women are white and it is Chef and Lance who grasp the opportunity to have sex with them. Chief Phillips willingly bows out in a moment full of racial poignancy while Clean hovers alternately outside the helicopter and the hut in which the two white men are having sex, but they ignore his pleas for inclusion. When one of the girls sees him through the window of the hut and cries out 'Who's that?' Clean forewarns us of his death, exclaiming 'I'm next, ma'am'.

Women are incidental to the plot of *ANR* and are not visible in the context of the prosecution of the conflict. However Bartlet's press secretary, C.J. Cregg (Allison Janney), is a woman (modeled on Clinton's Dee Dee Myers) and plays an important role in the daily affairs of the White House. C.J. is in regular face-to-face contact with the president and her opinion on key issues and decisions is often sought. As press secretary C.J. articulates

Bartlet administration policy for the media and public and filters and packages information for her male peers to enable them to protect the head of the household, the president. On this view, C.J. is at the cusp of the public and private domains and is charged with securing the domestic space of the White House by neutralising threatening issues before they reach the Oval Office. Nonetheless, the value of her job does not prevent her from being occasionally kept out of the loop by the central male characters, suggesting that she is excluded from the innermost workings of the administration. While Christina Lane takes a more positive view of the gender and race politics of *The West Wing* than we do, her reference to a scene in the episode 'Let Bartlet be Bartlet' in which campaign finance reform is at issue and which locates both C.J. and Charlie as 'essential reaction shots to the conversation' (between Bartlet, Josh and Leo) is revealing. Lane notes that C.J. and Charlie serve as visible reminders of the disenfranchised groups most affected by the problems of campaign finance in the US.[46] In this respect, C.J.'s (and Charlie's) marginalisation as a citizen is in part replicated in the Oval Office itself.

Moreover, other women in *The West Wing* are cast in peripheral roles.[47]

> Sure, the show says, women have a place in the White House. But they serve the boys: they don't initiate. The thinness of Kelly's character further isolates C.J. as both the only major female character on the show, and the only woman within Bartlet's kitchen cabinet. Her lack of female peers positions her as exceptional, an aberration from her sex, not its representative.[48]

In other words, the women serve as protectors and nurturers of their men. This applies to Mrs. Landingham (Kathryn Joosten) (whose sometime matronly tone with the president reflects a relationship that dates back to Bartlet's youth), Donna, with respect to Josh, and Margaret (NiCole Robinson), Leo's secretary. Interestingly, it is the female characters that do not work in the immediate orbit of the men of the inner circle – Mandy Hampton (Moira Kelly), Amy Gardner (Mary-Louise Parker), Ainsley Hayes (Emily Procter) and Joey Lucas (Marlee Matlin) – that most ruffle and challenge them. Their appearances generally coincide with some kind of crisis or problem that the men cannot themselves resolve, suggesting perhaps that like black soldiers in Vietnam, there are far more women involved in the heat of political battle than meet the eye. Nonetheless, having fulfilled their designated task the powerful women of *The West Wing* are nudged off centre-stage.

Ainsley, as a brash young neo-conservative, simply refuses to accept that her career could be limited by her gender and candidly opposes the socially progressive programmes advanced and defended by the younger men of the Bartlet White House.[49] But one wonders whether characters like Ainsley and Joey indeed serve to undermine affirmative action politics. Like Charlie, who is integrated into the White House because he is not politicised as a black, Ainsley and Californian Democrat Joey, both ambitious *and* attractive, have not let their gender, and in Lucas's case a disability as well, get in the way of successful careers. The nagging problem for liberals such as Sam and Josh is the lack of evidence affirmative action programmes have contributed to either of these political careers. Moreover, both women are accepted in and valuable to the White House precisely because they do not think like women, but like politicians. They make no demands on the part of women. Interestingly, both Sam (with Ainsley) and Josh (with Amy) have problems managing their relationships with powerful, independent women, suggesting they may well struggle should their idealised world of gender equality actually take shape.

Amy Gardener does, however, make explicit demands on the part of women:

> I spent a year and a half as issues director for NOW, two years as political director for Emily's List, founder of the Democratic Women's Forum, AA to Hope Schrader, and director of the Women's Leadership Coalition .... I beat back five anti-choice judicial nominees, got the Violence Against Women Act re-authorized with a Republican majority and raised $22 million for House candidates.[50]

On the surface, there is no question that Amy is ideologically closer to the Bartlet administration than she could ever be to a Republican government. Nonetheless, she is not in the administration's pocket, at least not before she is co-opted to serve as the First Lady's chief of staff, and continues to campaign vigorously on issues that she believes that the Bartlet White House neglects. It falls to Josh, who has a long-standing acquaintance with Amy, to attempt to neutralise problems that arise between 'women' and the administration. Josh despite his commitments to social justice is profoundly frustrated by what he sees as Amy's thanklessness and inability to see the broader picture. Moreover, his primary argumentative tactic is to highlight how much worse life for women would be under the Republicans, thus avoiding the merits of Amy's arguments. Even the romance which threatens to blossom is stymied by the political tensions that exist between them. Like Sam, Josh is unable to confront a powerful and independently minded woman on her own terms and clearly shows the (gendered) limits of his commitments to equality.

CONCLUSION: THE ETERNAL HOURGLASS

The United States is supposed to have lost its innocence in the jungles of Vietnam. On this view, the bright shining light on the hill of Kennedy's Camelot was extinguished by the war and the United States turned in upon itself in an agonising appraisal of the violence and corruption it revealed to itself and the rest of the world. Yet, remarkably, the contemporary American self that emerges from *The West Wing* is no more reflexive, no less naive and self-deluding than the one the United States supposedly lost in Vietnam. Indeed, perhaps the most vivid symbol of American moral decay, Richard Nixon, wrote of the post-Cold War United States that:

> American idealism – sometimes naïve, sometimes misguided, sometimes overzealous – has always been at the center of our foreign policy…This idealism has served as an indispensable foundation to sustain our commitment to the great moral causes of the twentieth century. It has enabled us to lead not on the basis of narrow and selfish interests but through the appeal of high ideals and common values.[51]

How did the United States regain its innocence?

The answer seems to us to be that the United States repeatedly proves itself capable of remarkable acts of forgetting. Notwithstanding the self-loathing that arose from its exploits in Vietnam, what never happened was the kind of reflexive examination necessary for transformation in the face of an innocence lost. Innocence is lost not with the realisation that the world is an unpleasant place, but with the acknowledgement that one's self or nation can contribute significantly to unpleasantness in that world. This is a lesson, at least by the time of *The West Wing*, the United States evidently has not learned, or more accurately has forgotten in the memory-washing Reagan era. While it was forgotten in and through films such as *Rambo*, the United States' capacity for forgetting is perhaps best represented by a more recent and considerably lighter film. In *50 First Dates*, Lucy (Drew Barrymore) has been injured in an accident and in a way that means she forgets her day every night. So each morning she awakes, thinking it is her father's birthday – the day on which she had her accident. During the film she repeatedly 'loses innocence', by being shown the clipping of the accident and having the previous year(s) explained to her, but the next morning she remembers nothing of it.[52]

The American capacity for forgetting is seen most clearly through the conscious expression of desire that is *The West Wing*. The vision is of the America that – liberals at least – would like to see: this is not a recognition of the state of the American self, but an expression of frustration that it is not as good as it could, and should be. We need think only of the mawkishly sentimentalised moments of American patriotism: the gang on the brownstone steps sequentially intoning God Bless America, or in the Oval Office, serving at the pleasure of the president of the United States, or the token Republican proclaiming the naives of *The West Wing* 'patriots'. Nietzsche's hourglass is rendered eternal in the United States through its seemingly endless capacity to forget its lost innocence.

Perhaps most poignantly, the continuing self-delusion of American innocence is seen in the special episode that precedes the third season. 'Isaac and Ishmael' (first broadcast 3 October 2001) confronts directly the questions asked in the aftermath of the terrorist attacks on Washington and New York three weeks earlier. As the title suggests, Aaron Sorkin tells a story of age-old hatreds – Isaac and Ishmael are the mythic originators of the Jews and Muslims respectively. While the episode works hard to distinguish clearly between Muslims and the 'extremists' responsible for the terrorist violence, it barely turns its gaze on the contemporary United States' place in the world. In this, of course, *The West Wing* is not alone, as the response to 9/11 has demonstrated American inability to consider the dark sides of its global practices. Nonetheless, the episode drew a huge viewing audience, some 25 million, perhaps indicative of the search for meaning and understanding in the weeks immediately after the attacks on New York and Washington. But if 'Isaac and Ishmael' is the high-water mark in terms of viewers, the episode was widely regarded as a critical failure, described, among other things, as pretentious and sermon-like. Arguably therefore, at the very height of its popularity, *The West Wing* contributed an important element to the collective sense of innocence wronged *and* misunderstands the *zeitgeist* of a post 9/11 America clamouring for vengeance. There seems little doubt that the politics of *The West Wing* are increasingly at odds with the United States of the Bush administration.[53]

The America that floats up the Nung River with Martin Sheen as he searches for Kurtz is an ugly, violently masculine self whose primary victims are the poor and/or dark-skinned. It is a self that cannot see the racist, gendered and class logic inherent in its violence: only the colour-blind eyes of Willard, Chef and Lance make it to Kurtz's own violently racialised dystopia. Reading *The West Wing* in these same terms reveals that the American self of even liberal fantasy at the end of the twentieth century remains wilfully blind to issues of race, gender and class and manifests itself in the boyish innocence of Sam and Josh. It revels in the eternal turning of the hourglass, secure in the belief of its innocence and goodness, certain it is still that bright, shining light upon the hill that draws to its shores the poor and huddled masses of the world.

ACKNOWLEDGEMENTS

An earlier version of this paper was presented to the YCISS Afternoon Seminar Series, York University, 13 November 2001. The authors would like to thank the participants for their thoughtful discussion. We would also like to thank the three anonymous reviewers whose insightful remarks and suggestions have helped us clarify our arguments.

NOTES

1. See Michael J. Shapiro, *Methods and Nations: Cultural Governance and the Indigenous Subject* (New York: Routledge 2004) pp.173–4.

2. Mark J. Lacy, 'War, Cinema and Moral Anxiety', *Alternatives* 28/5 (2003) p.614.

3. Ibid.

4. See inter alia C. Weber, *International Relations Theory: A Critical Introduction* (London: Routledge 2001), J. Weldes, 'Going Cultural: Star Trek, State Action, and Popular Culture', *Millennium* 28/1 (1999) pp.117–34, idem, 'Globalisation is Science Fiction', *Millennium* 30/3 (2001) pp.647–67, I. Neumann, ' "Grab a Phaser, Ambassador": Diplomacy in Star Trek', *Millennium* 30/3 (2001) pp.603–24, Mark J. Lacy, 'Cinema and Ecopolitics: Existence in the Jurassic Park', *Millennium* 30/3 (2001) pp.635–45; idem, 'War, Cinema, and Moral Anxiety', *Alternatives* 28/5 (2003) pp.611–36, Robert W. Gregg, *International Relations on Film* (Boulder, CO: Lynne Rienner 1998).

5. P.C. Rollins, 'United States–Vietnam Reconciliation in 1994: Feature Films on the Vietnam War', *National Forum* 74/4 (1994).

6. G. Kilday, 'The Two West Wings', *The Advocate*, 13 February 2001.

7. Ibid.

8. C. Lehmann, 'The Feel-Good Presidency – The Pseudo-politics of the West Wing', *The Atlantic Monthly* 287/3 (March 2001). See also Pamela Ezell, 'The Sincere Sorkin White House, or, the Importance of Seeming Earnest', in Peter C. Rollins and John E. O'Connor (eds), *The West Wing: The American Presidency as Television Drama* (Syracuse, NY: Syracuse University Press 2003) pp.159–60.

9. D. Saunders, 'Hollywood, D.C.', Weekend Review, *Globe and Mail*, 17 November 2001. See also Shapiro (note 1) p.199, Lacy (note 2) p.614.

10. Mark Poster, 'Theorizing Virtual Reality', in *The Information Subject* (Amsterdam: G+B Arts International 2001) pp.120–21.

11. Ibid. p.121.

12. O. Burkeman, 'Next week on The West Wing … erm', *The Guardian*, 5 May 2003.

13. Shapiro (note 2) pp.1–31.

14. See, for example, W. Safire, 'Essay: Syndrome Returns', *New York Times*, 30 April 2001, p.A19.

15. See, for example, Susan Jeffords, *Hard Bodies: Hollywood Masculinity in the Reagan Era* (New Brunswick, NJ: Rutgers University Press 1994) p.33. See K.J. Campbell, 'Once Burned, Twice Cautious: Explaining the Weinberger–Powell Doctrine', *Armed Forces & Society* 24/3 (1998) pp.357–75 for an account of the development of US military doctrine following the Vietnam War. Colin Gray (see C. Gray, 'The RMA and Intervention: A Sceptical View', *Contemporary Security Policy* 22/3 (2001) pp.52–65.), for example, disputes the possibilities for the RMA to achieve is zero-casualty objectives, but does not contest the genealogical point. For alternative readings of the so-called revolutions in military affairs, see, *inter alia*, James Der Derian, *Virtuous War: Mapping the Military–Industrial–Media-Entertainment Network* (Boulder, CO: Westview Press 2001) and A. Latham, 'Warfare Transformed: A Braudelian Perspective on the Revolution in Military Affairs', *European Journal of International Relations* 8/2 (2002) pp.231–66. Interestingly, Marlon Brando's Col. Walter Kurtz in *ANR* also recognises the limits of technology in attaining military superiority. In Kurtz's dossier, Sheen reads: 'Commitment and Counter-Insurgency, by Col. Walter E. Kurtz. As long as our officers and troups (sic) perform tours of duty limited to one year, they will remain dilettantes in war and tourists in Vietnam. As long as cold beer, hot food, rock and roll and all the other amenities remain the expected norm, our conduct of the war will gain only impotence. (In the document, but not read aloud – The wholesale and indiscriminate use of firepower will only increase the effectiveness of the enemy and strengthen their resolve to prove the superiority of an agrarian culture against the world's greatest technocracy … The central tragedy of our effort in this conflict has been the confusion of a sophisticated technology with human commitment. Our bombs may in time destroy the geography, but they will never win the war … We need fewer men, and better; if they were committed, this war could be won with a fourth of our present force…)' (see <http://www.filmsite.org/apoc3.htm>).

16. It is generally supposed that the Vietnam syndrome was systematically undermined during the Reagan administration, and then comprehensively overcome in the 1991 Gulf War. For a contrary argument, which suggests that the Gulf War in fact *reinforced* the syndrome, see Campbell (note 15).

17. J. O'Sullivan, 'A Sense of Dishonor? The case of Bob Kerrey', *National Review* 53/10, 28 May 2001.

18. Ibid.

19. R. Falk, '"The Vietnam Syndrome": the Kerrey Revelations Raise Anew Issues of Morality and Military Power', *The Nation* 273/2 (9 July 2001) pp.18–23.

20. The destruction is untold precisely because of the unwillingness of the United States to account for that destruction, and in particular to reckon the numbers of Iraqi dead. The independent tracking project, 'Iraq Body Count', estimated as of 29 July 2004 that between 13,086 and 15,149 Iraqi civilians have been killed in the war and occupation. (<http://www.iraqbodycount.net/>, accessed 7 October 2004.

21. Jeffords (note 15) p.38. Brando's unexpected corpulence compelled ANR director Francis Ford Coppola to film him in semi-darkness.

22. Ibid. p.13.

23. Ibid. pp.31–5.

24. Lynda E. Boose, 'Techno-muscularity and the "Boy Eternal": From the Quagmire to the Gulf', in M. Cooke and A. Woolacott (eds), *Gendering War Talk* (Princeton, NJ: Princeton University Press 1993).

25. Ibid., pp.80–87, 90.

26. Jeffords (note 15) pp.25–26, 138.

27. Ibid., pp.24–5.

28. Boose (note 24) p.75.

29. See Dan Glaister, 'The Governator II', *The Guardian*, 7 October 2004, G2, pp.2–3.

30. Boose (note 24) p.73.

31. Ibid. p.76.

32. Ibid. p.81

33. C. Fuchs, 'I'm Next Ma'am', *Poppolitics.com*, <www.poppolitics.com/articles/2001-08-13-redux.shtml>, accessed 19 September 2001.

34. Lacy (note 2) p.627.

35. Ibid.

36. There are blacks in senior positions in Bartlet's White House but none on his immediate staff. Admiral Percy Fitzwallace (John Amos) of the US Navy appears from time to time to advise Bartlet and is acutely conscious of his achievements as a black man. In one instance he reels off a list of accomplishments (in the context of his ethnicity) and invites those listening to 'beat that with a stick'. Nancy McNally (Anna Deavere Smith) is national security adviser to Bartlet. She is uncompromisingly tough, suggesting that an ambitious black woman cannot afford herself the luxury of Sam's and Josh's liberal moral angst. Both are listed as secondary characters at <http://westwing.bewarne.com/characters/secondary.html>.

37. H. Gray, 'Anxiety, Desire, and Conflict in the American Racial Imagination', in J. Lull and S. Hinerman (eds), *Media Scandals: Morality and Desire in the Popular Marketplace* (Cambridge: Polity Press 1997) p.96.

38. Robert G. Lee, *Orientals: Asian Americans in Popular Culture* (Philadelphia, PA: Temple University Press 1999) p.150.

39. Charlie confirms his salary in episode 19. See <http://westwing.bewarne.com/charlie.html>, accessed 19 August 2003.

40. According to Washington Trends, the average salary for 2001 was $37,864, or about $730 per week. See <http://www.ofm.wa.gov/trends/htm/fig102.htm>, accessed 19 August 2003.

41. See E. Miller, 'The Children of Light', *Christianity Today* 45/1 (1 October 2001).

42. Friedrich Nietzsche, *A Nietzsche Reader*, trans. R.J. Hollingdale (Harmondsworth: Penguin 1977) pp.250–51. Extract from *The Gay Science* (1882 and 1887).

43. Christina Lane, 'The White House Culture of Gender and Race in *The West Wing*: Insights from the Margins', in Peter C. Rollins and John E. O'Connor (eds), *The West Wing: The American Presidency as Television Drama* (Syracuse, NY: Syracuse University Press 2003) pp.34–5.

44. Kilday (note 6).

45. See Bonnie Honig, *Political Theory and the Displacement of Politics* (Ithaca, NY: Cornell University Press 1993) p.130.

46. Lane (note 42) p.34.

47. These include the president's secretary, Mrs (Delores) Landingham (Kathryn Joosten) (who dies), the vivaciously intelligent young Republican, Ainsley Hayes (Emily Procter), recruited by the administration to give them a heads-up on the thinking of the opposition (and who is accommodated in a basement office complete with water pipes and distant from the 'real' debates upstairs), Donna Moss, whose 'commonsense' marks her as a far more useful adviser on political pragmatics than Josh gives her

credit for, first season political strategist Mandy Hampton (Moira Kelly), whose grating advice and in-your-face manner alienate the male team, the deaf and politically astute Joey Lucas (Marlee Matlin) who acts as a political strategist and pollster for the Bartlet Democrats, Amy Gardner (Mary-Louise Parker), director of the Women's Leadership Coalition and latterly chief of staff to the First Lady, and Margaret (NiCole Robinson), the enigmatic secretary to Leo McGarry.

48. L. Smith, 'Reigning Men', *Pop Matters*, <http://www.popmatters.com/tv/reviews/w/west-wing.html>, accessed 19 September 2001.

49. In the episode '17 People' Ainsley declares: 'A new amendment we vote on declaring that I am equal under the law to a man? I'm mortified to discover there's reason to believe I wasn't before. I'm a citizen of this country. I'm not a special subset in need of your protection. I do not have to have my rights handed down to me by a bunch of old white men. The same Article 14 that protects you protects me.'

50. See <http://westwing.bewarne.com/amy.html>, accessed 19 August 2003.

51. Jeffords (note 15) pp.184–5.

52. *50 First Dates* is a more interesting representation of American forgetting than the rather better, and better known, film *Memento*. In the latter, Leonard (Guy Pearce) has suffered a similar brain injury to Lucy's, but the process of forgetting is much faster, and so he can forget everything in the time it takes for another character to leave a room and re-enter. Lucy retains her short term memories for the day, and so can reflect on her past, only to have the reflection wiped away by sleep.

53. See Ezell (note 8) pp.172–3.

The Frustrations of Geopolitics and the Pleasures of War: *Behind Enemy Lines* and American Geopolitical Culture

GEARÓID Ó TUATHAIL
Department of Government and International Affairs, Virginia Tech, Alexandria, Virginia, USA

In November 2001, Rupert Murdoch's Twentieth Century Fox Pictures rush released the movie *Behind Enemy Lines* to take advantage of a surge of American nationalism in wake of the 11 September ('9/11') attacks and the US-led retaliatory attack against the Taliban regime in Afghanistan. The military action adventure, loosely inspired by the shoot down of fighter pilot Scott O'Grady over Bosnia in May 1995, opened strongly, making almost $20 million on its first weekend of release. Over the subsequent months it made $59 million in domestic earnings and $7.1 million overseas, a healthy initial return on a $40 million budget production.[1] Video and DVD earnings are still accumulating.

The entrepreneur behind the production was John Davis of Davis Entertainment in Los Angeles, who wanted to make a 'patriotic action adventure' amidst the foreign policy crises of the 1990s. An initial story by John and Jim Thomas was written as a 'gung ho' buddy movie set against the backdrop of the Bosnian war. The project ran into difficulties, however, and was only revived after retired Navy pilots secured the cooperation of the Department of Defense.[2] Davis pitched the project to the US Navy as the next *Top Gun* (a film that greatly boosted Navy pilot recruitment) and reassembled that film's aerial photography crew for his film. Department of Defense cooperation meant access to some spectacular military hardware for the film. Two working aircraft carriers, the US *Constellation* and the *US Carl Vinson*, and a fleet of contemporary fighter jets and helicopters were made available to the producers.

Targeting an 'adolescent male' audience, Davis signed up a first time Irish director, John Moore, who had made a name for himself directing mini-movie fantasy commercials for Sega, Guinness and other corporate clients.[3] An initial screenplay by David Veloz and Zak Penn secured Gene Hackman to the project, a veteran not only of the US marines but of 'patriotic' Vietnam movies like *Uncommon Valor* (1983) and *Bat 21* (1988), also about pilots caught 'behind enemy lines'.[4] The somewhat unusual casting of the heretofore droll actor Owen Wilson in a heroic action role brought script revisions, with the storyline becoming, the film's promotional material explains, about 'a top naval aviator who is frustrated that fragile geo-politics have kept him from what he knows best: flying F/A-18 Superhornet jets into combat'.[5]

Behind Enemy Lines is a commercial product designed to make money.[6] Like many of News Corporation's product lines, it targets a large audience and panders to chauvinistic nationalism. The production process draws upon an archive of heroic filmic visualisations of the American military from the *Sands of Iowa Jima* to *Black Hawk Down* to create an updated product for the PlayStation age: the special effects are enhanced, the locations geopolitically intense, but the message is the same. The US military is an awesome and admirable military machine characterised by bureaucratic professionalism (modernity) yet peopled by distinctive personalities who, despite surface antagonisms, hold common values of duty, honour and country (tradition) which are at their finest in conditions of adversity. Out of a murky and ambivalent episode of the post-Cold War era, the US military's role during the Bosnian war, the movie invents a heroic moral engagement driven by a commitment to universal human rights. Like *Black Hawk Down* and many American Vietnam war films, this recovery of the heroic amidst a historical engagement that was far from it is a striking feature of the film and one of the reasons why I find it worthy of examination. A second reason is the reliance of the film on a familiar yet nevertheless eyebrow-raising story line that casts the US's own NATO allies as bureaucratic obstacles to heroic moral action. A third reason why I find the film of interest is that it is a window into the contemporary culture of US militarism, a problematic that requires urgent examination and analysis given the way the administration of George W. Bush has used 9/11 to invade and overthrow governments in two states, and dramatically increase what were already the largest levels of defence spending by any state in the world.

There are many other aspects of *Behind Enemy Lines* worthy of investigation: its gendered politics, its pornographic gaze on weaponry, its military-industrial-entertainment qualities and the historical resonance of its storyline.[7] Developing out of the points of interest above, I examine three themes that are embedded in the film but that open out on to much larger questions about American geopolitical culture and the American military. The first theme is conceptual and examines the movie's story line as an articulation and reproduction of what Mead terms the 'Jacksonian' geopolitical tradition in American geopolitical culture.[8] The second theme is historical and considers the movie in relationship to the Scott O'Grady story and the historical record of the US military during the Bosnian war. The final theme examines the movie as an articulation and visualisation of the structure of feeling of American militarism. One striking features of the marketing of *Behind Enemy Lines* is its use of 9/11 imagery. A promotional poster for the film features the abandoned pilot charging forth, pistol in hand, firing back in revengeful anger against the enemy stalking him *against a background of two towers of light surrounded by clouds of smoke.*[9] The image is iconic of post-9/11 America. The superpower was roused, angry and shooting out blindly at those stalking it. A revengeful America was unleashed and 'pay back' was about to begin.

A JACKSONIAN STORYLINE

The beginning of *Behind Enemy Lines* offers viewers the military equivalent of a failed erection. With rock music pumping and planes coming and going, combat navy pilots Stackhouse and Chris Burnett (Wilson's character) climb into the cockpit of their F-18, lock and load for takeoff, are boosted for a catapult launch off the aircraft carrier and then, at the last minute, are taken down off their boost as their mission is cancelled. 'Are you kidding me? That's happened twice today', declares an angry Burnett. The ensuing dialogue expresses their frustration at not being able to fly and see some combat action. Burnett ('burn-out'), a once promising naval aviator, is shown impotently eating jello as he explains how he has lost his desire to stay in the Navy. Called before his commander, Admiral Leslie Reigart (played by Gene Hackman), he explains why he no longer wants to 'serve his country': 'I signed up to be a fighter pilot. I did not want to be a cop. I certainly did not want to be a cop in a neighborhood nobody cares about. That and the routine in the ship kind of wore me out.'

Responding to Reigart's claim that 'we are at war' he responds:

> If we are at war why doesn't somebody act like we are at war because as far as I can tell we go out, we fly around and we come back. Now, maybe we're pretending we're in the middle of a fight but that is all we're doing. We're pretending. Because we're not fighting. We're watching.

The condition of Burnett is a metaphor for the US military in the post-Cold War world. The moral clarity of the past and the sense of mission that came with it are no more. Burnett establishes this by stating that 'everybody who signed up wants to punch a Nazi in the face at Normandy but those days are long gone'. There are no more enemies anymore, just conflicts in regions that are not significant. And, rather than settling these conflicts, America is on the sidelines watching them and pursuing a confused, directionless policy. As Burnett perceives it, one day America is aiding one group, the next day another. This is the condition of the 'new world disorder' and Burnett does not like its lack of meaning nor America's inconsistent purpose and unclear mission.

The rest of the movie is about Burnett's rediscovery of a sense of meaning and mission as a combat navigator. Eventually getting airborne, he deviates from his scheduled flight path and discovers something meaningful (the location of a mass grave). But suddenly he is shot down by a Bosnian Serb surface-to-air missile and his surviving co-pilot is executed. He is now the subject of a manhunt behind what have clearly become *enemy* lines. In the course of avoiding his pursuers, Burnett is renewed as a patriotic American military professional as he demonstrates courage, determination and initiative in avoiding capture. We experience the adrenaline rush of battle and the pleasure of a final 'cavalry coming' shoot-out scene. The emotional climax of the film sees him reunited with his military brothers and father-figure Reigart, crumpling up an earlier letter of resignation from the Navy. He has renewed his sense of meaning and mission as he returns with photographic evidence that will be used to convict the war criminal that pursued him.

As a film about remasculinisation through the transcendence of trial and tribulation, *Behind Enemy Lines* has certain similarities with the Vietnam revenge films of the Reagan era like the *First Blood/Rambo* series, the *Missing in Action* series, and *Uncommon Valor* (1983).[10] In her study of gender in popular representations of the American military experience in Vietnam, Jeffords argues that in these and other cultural productions 'war is the spectacle of the masculine bond. It is the optimal display of masculine collectivity in America, since battle, as defined by American culture, is an exclusively male activity.' 'War', she writes, 'is the stage where men can perform, for themselves and their "enemies", it is the ostentation of bonding.'[11] This theme is undoubtedly crucial to *Behind Enemy Lines* where a father–son relationship between Reigart and Burnett is explicitly signaled throughout ('Hold on, son, we're coming to get you').

While remasculinisation and the spectacle of the male bond are central in *Behind Enemy Lines*, these themes are broad ones that characterise many American war films (from *The Sands of Iwo Jima* to *Independence Day*). Jeffords argues that gender rather than patriotism and militarism is the primary impetus driving the narrative of films like *Rambo* but it is important to not loose sight of these works as geopolitical storylines and visualisations that are part of larger geopolitical traditions and culture.[12] These films construct a world of insides and outsides, friends and enemies, and the danger that results in transgressing certain established, albeit unjust, lines of power. Remasculinisation proceeds through transgression of these rules and lines of power. Enemies are recognised, confronted and finally defeated. But, interestingly, these 'enemies' are not solely foreign outsiders who constitute the official geopolitical adversaries of the state. There is also a second enemy, an enemy within, a domestic bureaucrat or malevolent agent who tries to thwart the hero from achieving his destiny. A striking feature of so-called 'patriotic' Vietnam revenge films is that the enemy is often an agency or employee of the United States government. In the discourse on the Vietnam war favoured by many right-wing veterans, these are the agents of a feminised and internationalised state, 'Washington bureaucrats' who refused to win the war and forced ordinary soldiers to fight 'with one arm tied behind their back'. For Jeffords, the association of loss with the government and the honour of war with the solider is a principal thematic of the discourse she examines.[13]

Behind Enemy Lines articulates its own particular post-Cold War version of this discourse. Its storyline contains the rather remarkable contention that NATO, the European alliance system established and dominated by the United States, is a major obstacle to Burnett being rescued. The central antagonism is perhaps meant to be read more as one between the United States and France for the alliance is personified in the film by a French Admiral Piquet (played by the Portuguese actor Joaquim de Alamedia). Piquet is Reigart's superior commander and he demonstrates more concern about the fragile state of diplomacy than with rescuing a fallen American pilot. The rescue attempt he authorises, an expedition by French marines, is aborted just as it approached the rescue of Burnett. When Reigart establishes that Burnett is still alive, he authorises and leads a rescue mission himself, deliberately not informing Piquet and thus violating the chain of command. For the renegade rescue of 'his boy', Reigart looses his command. The film concludes with a titled end story that he 'was reassigned to an "administrative" job in Washington' (the scare quotes and location evoke the pejorative "government bureaucrat"). But there is a satisfying twist: 'He chose instead to retire with the respect and gratitude of the men and women (sic) under his command.'[14]

This antinomy between Reigart and Piquet is symbolic of a much larger set of conflicting principles: war versus geopolitics, action versus talk, the uncomplicated and direct versus the complex and bureaucratic, unilateralism versus multilateralism, war-fighting versus internationalised diplomatic peacekeeping (see Table 1). In keeping with a long tradition in American geopolitical culture, *Behind Enemy Lines* is another parable about transgressive direct action cutting through the Gordian knot of a problem, about the liberation of a war-fighting masculinity from the constraints of multilateralism and diplomacy in order to 'get the job done'. It is the story of a renegade Rambo fighting his own personal war with higher values than that of the government. It is Lieutenant Colonel Oliver North's storyline during the Iran-Contra scandal. It is an acceptable 'breaking of the rules' of the institution or society in the service of a 'higher good', a more noble truth. In fact, the storyline is an exaggerated expression of the very values and principles represented by the military: duty, honour, country.

Within American geopolitical culture, this story line is closely associated with the geopolitical tradition Walter Russell Mead names after General and later President Andrew Jackson: the Jacksonian tradition. This tradition is the least intellectual of the four traditions he identifies yet, he argues, it is an expression of the social, cultural and religious values of a large portion of the American public. The Jacksonian tradition is America's warrior culture,

TABLE 1 The Jacksonian Storyline as a 'War' versus 'Geopolitics' Contrast

	War	Geopolitics
Form of military masculinity	Heartland military masculinity; 'one of the guys'. Hetrosexual.	Internationalised military bureaucrat Not fully heterosexual.
Military command style	An 'uncomplicated man' who is a father figure to 'his men.'	An aloof, snobbish, internationalised, intellectual commander.
Mode of operation	'Straight shooter'. Favours direct military action. War fighting. Unilateralism.	Negotiation. Follows diplomatic rules and conventions. Peacekeeping. Multilateralism.
Highest Value	Respect, professionalism, and loyalty to one's own immediate military family.	Political success through compromise and manoeuvre.
Conception of Military	Military as essence of 'the nation', as representing selfless service to country. Needs to act independently sometimes to serve a higher goal.	Military as an instrument of political leaders of the nation. Constitutionally subservient.
Desire	To be unrestrained. To fight against clear enemies. Righteous violence.	To be diplomatically successful. To bring peace to the region. Pragmatism.
Behind Enemy Lines characters	Chris Burnett (Wilson); Admiral Leslie Reigart (Hackman)	Admiral Miguel Piquet (NATO)
Community	The organic community of the men (and women) abroad aircraft carriers; 'America'.	The European states that comprise NATO. European capitals. 'Washington'.
Real life example	Admiral Leyton Smith.	General Wesley Clark. (Richard Holbrooke)
Presidential exemplar	George W. Bush and his unwavering and resolute leadership ('You know what I believe').	John Kerry ('looks French') and his 'global test' before US force can be used.
Policy priority	Pre-emptive war fighting	Multilateral diplomacy and international peacekeeping.

a tradition that emphasises and celebrates American military power. Strongest in 'red state' America (i.e the states that voted for George W. Bush for President), the US military is seen as the embodiment of the nation as an idealised folk community united by shared values and a common destiny. Heavily mythological, these values centre on notions of respect and honour. Serving one's country is the noblest of activities and callings. An honourable person is a person who is ready and willing to die in the service of 'his' country.[15] Those who questions this system of honour and respect, as Burnett does before Reigart at the beginning of *Behind Enemy Lines*, are held in barely disguised contempt, as once promising sons who have lost their way.

The drama at the centre of *Behind Enemy Lines* is of a prodigal son finding his way back to the folk community that drops everything to rescue him. Along the way, he reveals himself as a worthy member of that community. He demonstrates considerable self-reliance when his plane is shot down and he is forced on the run. His individualist enterprise is apparent as he negotiates a series of obstacles and challenges from forces tracking him. Forced to rely on the training provided to him by the military, preparation he had previously questioned, he reveals initiative, courage and considerable close-quarters martial skill as he triumphs over the tracker who had pursued him with dogged determination. Faith and belief in the bedrock values of the military, and of the Jacksonian tradition more generally, are restored.

THE BOSNIAN WAR AND THE US MILITARY

Behind Enemy Lines is one of only a few American financed productions to engage the subject matter of the Bosnian conflict. At the outset, Burnett seems to articulate a critique of the American military role in the conflict. 'We're not fighting. We're watching.' Yet it is more the former than the latter that concerns Burnett. The American military is tied down by restrictive rules of engagement and circumscribed by diplomatic considerations and fragile geopolitics that emanate, it is implied, from America's European alliance partners. What this means for Burnett is that flight training is cancelled and, when he does get to fly, there are restrictions on what flight path he can and cannot take. As an American military 'everyman', Burnett expresses frustration with these restrictions, with the oscillating character of US diplomacy, and the general confusing nature of the Bosnian war. In the early jello-eating sequence of dialogue at the outset of the film he declares:

> At least give me a fight I can understand. Good luck talking to someone at home, trying to explain: 'well, today's Tuesday, I think we are helping these people. No, no, now its switched around, now we're helping … (voice falls off in disgust). Its, like, a joke.

Safely distant from the Bosnian war on an aircraft carrier in the Adriatic, Burnett is shot from the cocoon of his F-18 and is suddenly face to face with the horrors of the war as he stumbles into an open mass grave site. A *Schindler's List*-style black and white flashback scene makes clear the history of the site to the audience. The war is rendered black and white – perpetrators and victims – and Burnett appears to get a fight he can understand and explain to someone at home. What appeared confusing on the ship is not on the ground: a Second World War-like Holocaust with clear moral divides. What this makes possible is the realisation of an already articulated desire to 'punch some Nazi in the face'. Liberated from geopolitical constrictions and confining rules of engagement, the Bosnian war makes possible the pleasure of moral supremacy, and the consequent thrill of administrating righteous violence. Burnett has found the world he wants, a friend and enemy world where he has a mission and righteous desire for revenge. Encountering a Muslim resistance leader later in the film, Burnett declares: 'I am an American. I am on your side.' The Muslim leader is less than impressed, though, revealing hostility and anger as he manhandles Burnett and declares 'American government for shit!' His actions reinscribe him in the world of Balkan politics, as just another of the 'warring parties' rather than a grateful ally. But America's moral virtue does not need to be recognised even by those it would help. Burnett reveals that America is on the side of international justice as he successfully rescues photographic evidence of the mass graves the Serbs were trying to hide. With this evidence in hand, the international community is able to uncover the site of the mass grave and arrest the Serb leader Lokar (played by Olek Krupa), the figure directing the pursuit of Burnett, for war crimes.

The final released version of the film does not show the excavation of the mass grave – a shot that ends with a fluttering United Nations flag – nor the arrest of Lokar by international police, both of which had been filmed by the director. The last few brief shots deal only with Reigart's 'respect and gratitude' retirement. Audience screen tests revealed that the emotional climax of the film was over for the audience and they wanted to leave. And so the movie gives them what they want: a film that is not about the Bosnian war or an American pilot as UN war crime evidence gathering hero. Rather, the Bosnian war is a conveniently polarised moral backdrop to celebrate the US military and reaffirm it as a righteous force even in a confusing and ungrateful world. *Behind Enemy Lines* is not simply about the rescue of a downed American pilot in Bosnia but about the rescue of a lost American military image and mission amidst a confusing 'new world disorder'.

Propaganda can be defined as works of mass media deliberately designed by powerful state institutions to condition popular geopolitical attitudes and promote loyalty to the state, particularly the military. Defined as such, *Behind Enemy Lines* is a propaganda film. It is an officially approved mass media mythologisation of the US military's role in Bosnia.[16] The record of the US military in Bosnia until August 1995 was more than simply 'watching'. The US military helped structure the conditions under which the war was fought. It is an irony of the US military involvement in the Bosnian war that the very desire to avoid the perceived limited nature of the Vietnam war lead the US military into a highly restricted engagement with the Bosnian war.

A particular 'lessons of Vietnam' script informed the development of the US doctrine governing the use of military force. Galvanised by the ill-fated intervention of US forces in Beirut in 1983, this doctrine was formally codified as the Weinberger and later the Powell Doctrine. It held that US military force should not be used unless the US government had a clear political objective, was prepared to use overwhelming force to achieve victory, and had already determined for itself an acceptable 'exit strategy'. Summarised crudely, the doctrine held that 'without a will, a means to win and a way out, do not go in'.

The hegemony of this doctrine rendered the possibility of using military force to advance US foreign policy extremely rigid. The use of force was 'all or nothing' and, as articulated by Powell, it deprived political leaders of the flexibility of using limited and circumscribed force – e.g. once off air strikes – to achieve very localised and particular ends.

The rigidity that the Powell Doctrine institutionalised at the heart of US foreign policy gave rise to an equally rigid conception of US national interests and foreign policy discourse. Conflicts like the Bosnian war either involved US national interests or they did not. The conflict was either a simple case of international aggression by a large state against a smaller state (the Kuwait analogy) or it was a complex and intractable civil war (the Lebanon or Northern Ireland analogies). There was little room for policy nuance and for understanding the Bosnian war on its own terms. Under the chairmanship of General Colin Powell, the Joint Chiefs of the US military decided that the breakup of Yugoslavia and the subsequent war in Bosnia did not involve any serious US national interest. In coming to this conclusion, Powell was in harmony with the prevailing geopolitical wisdom in the Bush administration, especially Secretary of State James Baker, who had famously quipped that 'we don't have a dog in this fight'.[17]

But Powell and the US military were a powerful institutional force with their own agenda which was, as they understood it, to preserve the power and the readiness of the US military to 'defend the nation' from existential threats. Powell and nearly all of his colleagues concluded that Yugoslavia was a potential military quagmire that did not merit US involvement. It was a potential Vietnam but this time the US military would stand up to the politicians and force them to confront hard political and military questions before they deployed the US military. As Powell saw it, US politicians had a tendency to establish an American 'presence' in 'trouble spots' without a defined mission, as a 'symbol' rather than as part of a national security strategy.[18] During the 1991 Serb-Croat war as Serb forces were directing artillery barrages against the city of Dubrovnik, there was a window of opportunity to conduct air strikes against Serb artillery positions and signal to their leadership that they could not do as they pleased to largely defenceless cities and civilians. The US Air Force Chief, General Tony McPeak, was in favour of air strikes and argued that they could be effective. Powell and NATO Supreme Allied Commander General John Galvin (the SACEUR is always an American), however, overruled McPeak and refused to intervene. In his account of US military thinking at the time, Halberstam recounts Galvin's thinking in which the quagmire metaphor was central:

> Like Powell, he (Galvin) believed ... that you could not just put your toe in ... if you put in your toe at Dubrovnik, the foot would inevitably follow, and if the foot was in, the rest of the body might also be pulled in.

> If we went in, who was with us? How far would our allies go? Where was the Congress and where would the media be?[19]

Note that there is no actual analysis of the Bosnian war in this reasoning. The metaphor overdetermines the conflict. The US military leadership, using reasoning like Galvin's, fought hard to avoid intervention in the Bosnian war. When Bosnian Serb concentration camps were uncovered in August 1992 and it was clear to most that the Bosnian Serbs were engaged in war crimes against the Bosnian Muslims, Powell argued forcefully within the administration that the US military should stay clear of what he represented as a Balkan 'civil war'. 'In Bosnia, we are dealing with an ethnic tangle with roots reaching back a thousand years.'[20] Where others saw genocide and clear victims and perpetrators in Bosnia, the Pentagon leadership and the top civilian leadership in the Bush administration saw, in the president's words, a 'blood feud' and 'complex, convoluted conflict' between Balkan peoples over 'century-old feuds' and 'age-old animosities'.[21] The storyline was patently and demonstrably false but the Pentagon and the Bush administration pushed it because it served their interests of not getting engaged in Bosnia during an election year and not allowing the US military to be dragged into a war in what they had constructed as a strategically marginal region.

But the Bosnian war did not go away. The US military was involved whether it liked it or not. The United Nations Security Council (UNSC) passed an arms embargo against the region which had the effect of consolidating the military superiority of Serb forces across the region. In enforcing this with their NATO allies, the US military was thus by the middle of 1992 directly maintaining two conditions that favoured the Serbs. First, they were promoting the very same story line on the war as that promoted by the ethnonationalist perpetrators, namely that it was an inevitable ethnic and religious war between three essentialised peoples who have hated each other for 'thousands of years'. The Bosnian Serb leader Radovan Karadžić returned the favour echoing the situation descriptions of the Pentagon and Foreign Office of the conflict as the 'next Vietnam' and an intractable Northern Ireland-like conflict.[22] Second, NATO and its allies were prohibiting the victims from defending themselves in the name of opposing any 'level killing field'.[23] What this did, of course, was make the 'killing field' one sided, turning a potential war into slaughter.[24]

The weakness of the Bush administration's policy towards the Bosnian war became an issue in the US presidential election. Candidate Clinton called for greater use of force in the region to protect relief supplies though he never challenged the humanitarian reading of the conflict favoured by the Bush administration. The shoot down of an Italian cargo plane delivering relief supplies to Sarajevo in September, media exposes of the use of cluster bombs by Serb planes against defenceless population centres, and Clinton's attack all added pressure on President Bush, to which he responded by supporting a ban on Serbian combat aircraft over Bosnia and Herzegovina in early October. The policy change was strenuously debated in the administration, with General Powell publicly questioning the need for a ban. Ostensibly, the ban would establish a new threshold in which NATO would be acting to protect Bosnian Muslims from air attack.[25] Powell argued that this was a 'slippery slope' leading to deeper involvement in the region.[26] The British and French governments were also leery of the policy change.[27] Powell went on the offensive in the media writing an opinion editorial in which he described the war in Bosnia as 'especially complex' with 'deep ethnic and religious roots that go back a thousand years'. The use of limited military force would not resolve it.[28]

The strenuous lobbying efforts of the US military and its overseas allies worked, for when the UNSC passed a resolution banning combat flights over Bosnia there was no enforcement provision attached. Instead the resolution stated that if the Serbs continue to fly military planes, the Security Council will 'consider urgently the further measures necessary to enforce this ban'.[29] Predictably, the flight ban was violated by the Serbs. In early December, the US government confirmed more than 200 unauthorised flights in Bosnia by fixed-wing warplanes, helicopter transports and gunships.[30] Whether the Serbs were still bombing and strafing Bosnian villages and towns was a matter of dispute.[31]

It was not until 31 March 1992 that the United Nations Security Council finally passed a flight ban with an enforcement mechanism. UNSC Resolution 816 authorised NATO airplanes to shoot down violators of the flight ban over Bosnia-Herzegovina.[32] In its first 'out of area' mission over territory of a nonmember and in support of the United Nations, NATO launched Operation Deny Flight in mid-April with the participation of American, British, French and Dutch airplanes.[33] The mission was one that NATO commanders, particularly the Americans, did not like. Having doggedly fought involvement in the Bosnian war and opposed enforcing a 'no-fly' ban, they were now involved in the war. Worse still, their involvement was a deeply circumscribed and limited one. The flight ban, for example, was over Bosnian airspace but not over Croatian airspace. To the Serbs seeking to create a greater Serbia on the ground, the border between the Bosnian and Croatian krajina did not matter. In northwest Bosnia, they moved their air operations to the Udbina airfield in Croatia, some 30 kilometers south-southwest of Bihać. Further, the US military found itself in the middle of a deeply political and multilateral war whose ultimate expression was the development of a 'dual-key' authorisation system for the use of force to protect UN operations in Bosnia.

NATO aircraft saw action over Bosnia in February 1994 as the Serbs violated the no-flight ban to bomb a Bosnian army munitions plant in Novi Travnik. Four of the six Serb aircraft that participated in the raid were shot down. NATO aircraft subsequently conducted 'pin prick' attacks against

Bosnian Serb ground targets in April, August and September. A military offensive by the Bosnian Army V Corp in Bihać prompted Serbs aircraft from Udbina to bomb the town of Bihać with napalm and cluster bombs. NATO responded by bombing the runway at Udbina to 'send a signal' to the Serbs not to violate the combat flight ban over Bosnia.[34]

It is clear that US commanders were not happy with the position they found themselves in off the coast of Bosnia. Their mission was to support 'peacekeeping', yet it sometimes required the use of force which necessitated a logic of war for robust force protection. They chafed at the 'dual-key' restrictions and diplomatic restraints on their rules of engagement. One of these commanders was Admiral Leighton Smith, chief of US Naval Forces in Europe. After November 1994, some missiles had been launched from Udbina at US aircraft participating in Deny Flight. Smith had to ask permission from General Bertrand DeLaprelle, UN commander at the time, to go after any missile sites and radars that threatened NATO aircraft. His request was denied by DeLaprelle, who reemphasised their peacekeeping mission.[35] Smith was not happy and made his views known. After November, NATO aircraft were faced with increased surface-to-air missile radar 'paintings' and had to dodge missiles on a number of occasions. Military logic dictated that Smith engage in suppression of enemy air defences for Deny Flight missions but diplomatic conditions, with UN hostages on the ground, made that provocative and dangerous.[36]

Another commander was Lt General Michael Ryan, commander of NATO Air South and the US 16th Air Force. Working directly with Smith since September 1994, his frustration at their situation was strong:

> All of us who had been in Vietnam were affected by the restrictions in Bosnia. The circumstances under which we had to operate in Bosnia with respect to surface-to-air defence reminded me of Vietnam to a fare-thee-well. In Bosnia we were not allowed, or there was no political authority or will, to go after the air defense systems, yet we wanted to make political statements with our bombing. Those kinds of restrictions are whacko. That is a stupid way to run a war. People say, 'This isn't war, it's a peacekeeping situation.' It's a war in the air here, I guarantee you.[37]

For American Air Force commanders used to overwhelming force, air supremacy and robust rules of engagement set by the US commanders alone, Bosnia was a difficult exercise in multilateral military decision making within a larger dynamic geopolitical context. Many experienced it as 'déjà vu all over again' (like Vietnam) and as a 'diminution of US power'.[38] Military antagonism between the Serbs and NATO reached a new high in late May 1995 when Serb forces retaliated against a NATO attack on a Serb ammunitions dump in Pale by taking more than three hundred UNPROFOR 'peacekeeping' soldiers hostage.

On 2 June 1995, a US F-16 piloted by Captain Scott O'Grady on a routine Deny Flight mission was shot down by a surface-to-air missile in northwest Bosnia. O'Grady managed to eject from the aircraft and parachute safely to the ground. Blown a considerable distance from the crash site he concealed himself in the forest for days, rarely moving after the second day.[39] The prospect of a US pilot lost in Bosnia was always going to test the degree to which American commanders could restrain their unilateralist instincts. Smith's first reaction was blunt. He told the NATO liaison officer to inform the UNPROFOR commander General Janvier in Zagreb that 'We've had an F-16 shot down, I'm coming to get him. When I find him, I'm going to pick him up. Don't anybody get in my way because if you do, I will kill you.' The same message was to be delivered to the Serbs, with the addition that 'I'm not kidding and I'm not asking permission.'[40] Though he had members of his own forces held captive by the Bosnian Serbs at the time, General Bernard Janvier was 'completely cooperative' on the process of agreeing to a combat rescue mission. He put pressure on the Bosnian Serb General Mladic to return the pilot alive.

It was five days before NATO established radio contact with O'Grady and determined that he was alive and not in captivity. Smith quickly established a rescue force consisting of no less than 40 aircraft and 60 US marines aboard the US *Kearsarge* in the Adriatic. At the last minute the rescue operation was re-designated as a NATO and not a solely US mission. Since each had different rules of engagement – NATO missions do not permit 'preemptive fire' – this caused some confusion but Smith made it clear what philosophy predominated: 'If you even think somebody's going to shoot at you, take 'em out.'[41] The mission successfully extracted O'Grady, with some helicopters dodging SA-6 missile fire on the egress. 'When they came back', one of the crew on the *Kearsarge* recalled, 'it was like that scene from *Top Gun*.'[42] Clearly, a movie was already imagined.

Behind Enemy Lines is not the Scott O'Grady story but its story line does capture some of the frustrations and unilateralist instincts of the US military forces operating off Bosnia.[43] The conflict the movie depicts between a NATO commander and a US commander did not occur but frustrations with the encumbrances of 'geopolitics' were widespread (though the US military helped create these conditions). A narrow self-absorbed moral vision predominated. In portraying Bosnia as a mere backdrop to US military dramas, the film unintentionally conveys this larger moral indifference to the Bosnian war (even as it seeks to recuperate the moral image of the US military). As we now know, some if not most members of the US military leadership considered the warring Bosnian parties moral equivalents and held all in contempt.[44] US civilian and military leaders devoted considerable resources to rescuing their own fighter pilot from the region yet stood by and watched as the largest massacre of civilians in Europe since the Second World War unfolded in Srebrenica.[45] Figures like Admiral

Leighton Smith, a possible inspiration for the Reigart character, were indifferent to the outcome of the Bosnian war, arguing that Bosnia was not worth US involvement in the first place.[46] Yet, he became commander of the Implementation Force (IFOR) charged with enforcing the Dayton Peace Accords, and pursued a narrow 'minimalist' role that placed force protection above political goals in the country. The Pentagon leadership refused to have IFOR adopt any kind of policing role or aid the civilian mission to implement Dayton.[47] It refused to arrest war criminals and stop Serb thugs from burning the Serb-held suburbs of Sarajevo in a final act of criminal destruction as Dayton took hold.[48] Many US politicians and military commanders simply wanted to end the war, stabilise the region, and get out.[49]

Given this far from glorious record, *Behind Enemy Lines,* like *Black Hawk Down,* is a cinematic rescue mission for the United States' military image, plucking a fictional moment of American virtue from the Bosnian miasma. The fantasy is that while a NATO-encumbered United States was merely 'watching' and not taking sides, a more authentic unencumbered United States was taking moral action, even if it broke some diplomatic rules. Though unilateral and narrowly self-serving, this America actually serves universal goals for it secures evidence that speeds the process of international justice. However, in the post-9/11 world, this end is cut from the picture and we are left with a final image of a resolute 'go-it-alone' leader (Reigart) being hailed by the personnel of an American aircraft carrier.

BEHIND ENEMY LINES AND THE 'NEW MILITARISM'

Before its theatrical release, *Behind Enemy Lines* was first shown on the deck of *USS Carl Vinson*, the aircraft carrier that features in the film. It was then officially premiered aboard the *USS Nimitz* off the coast of California, with the media and stars specially flown in for the event.[50] Producer John Davis described the scene as 'very emotional', while director Moore spoke of how great it was to have support at the highest levels of the Department of Defense. Working on the movie, he said, 'gave us the chance to interact with the 'guys' (sic) the pilots, mechanics and technicians. Their help and input were phenomenal. I made the movie for them.'[51]

At that moment, 'the guys' and the American state were at war, a world-wide war against terrorism declared by President Bush in the wake of the 9/11 attacks. A little over two years later, after brief but bloody invasions in Afghanistan and Iraq, the president would use a similar aircraft carrier off the coast of California for his own cinematic close up, flying solo onto the specially positioned carrier and emerging as a triumphant fighter pilot president. After a quick change of gear, he got behind the presidential lectern and beneath the banner 'Mission Accomplished' declared that major combat operations in Iraq had ended but the world-wide war against terrorism would go on indefinitely.[52]

Envisioned as a post-Cold War action adventure, *Behind Enemy Lines* was potentially a film product out of sync within the transformed geopolitical context after 11 September. That it actually appeared in synch with the times and rode a wave of nationalism to box office success is a testimony to how it successfully projected themes and emotions of longstanding resonance in American film and popular geopolitical culture. It offered a vision of an America adrift until a sudden crisis tests its character and provides meaning and a mission. It projects an America that is not aggressive but when attacked responds with righteous violence. It represents America as a unilateralist power yet also one that serves the greater cause of international justice. These three aspects of the film provide an interesting window into the structure of feeling of what Michael Mann has termed the 'new militarism' of post-9/11 America.[53]

First, *Behind Enemy Lines* articulates an 'everyman' frustration with the confusion of the post-Cold War era and then delivers on the desire for a Manichean world of clarity and moral certainty. Action and righteous violence are made possible by a clarified world where there are recognisably othered enemies and adversaries. The United States military is the instrument of moral righteousness and international justice. Post-9/11 geopolitics was constructed in such a manner by the Bush administration. The attacks were interpreted as acts of war rather than acts of criminality and terrorism. Terrorism became the equivalent of war. The struggle against the perpetrators was inflated into a global struggle against terrorism. Rather than being a localised threat from a criminal conspiracy, Al Qaeda become a 60-state conspiracy and only one of a myriad of terrorist organisations that the Bush administration decided were at war with the United States.[54] The conscious articulation of a Bush doctrine forced the world geopolitical map into Manichean categories of 'us' versus 'them'. States were either 'with us' or 'with the terrorists'.

The Bush administration's response to 9/11 marked the downscaling of the world of geopolitics into the categories of war. A sharp line was drawn between those inside the folk nation and those beyond it who dwell in the realm of darkness, hatred and distorted morality. The essence of the folk nation is the US military, who are seen as representative of the 'heartland values' of the nation: duty, honour, country. They are the defenders of 'freedom' and the instrument of the 'liberation' of peoples from tyranny and terrorism.[55] The enemy they face is technologically modern and determinedly anti-American, relentlessly driven yet also quintessentially other with no moral compass and driven by hate. In this fight, the prime responsibility of the folk nation is to itself alone. While there is much Wilsonian rhetoric about 'liberating' foreign peoples from oppressive regimes, the US military devotes its considerable resources and energies, first and foremost, to 'force protection'. Reducing 'collateral damage' is a priority that pales in comparison. In this sense *Behind Enemy Lines* accurately renders the narrowly self-interested and superficial nature of the US engagement with the places where it intervenes. Understanding the political dramas of these places is secondary to US military priorities and preoccupations. This is militarism in that it allows Jacksonian story lines and US military imperatives to drive America's geopolitical engagement with particular states and regions. For Michael Mann, the 'new militarism' is the use of 'militarism offensively to remake the world into a better place' which he contrasts to more traditional Cold War 'mainly pragmatic and defensive notion of military power'.[56] But this defines the new militarism too much in terms of Wilsonian story lines to the neglect of Jacksonian story lines of revenge. It is the overwhelming of the frustrations of geopolitics by the pleasures of war that is, I would argue, central to the structure of feeling of contemporary American militarism.

Second, *Behind Enemy Lines* celebrates the impulse for unilateralist action on the part of the United States and impatience with the pragmatics of geopolitics. The conflict between Reigart and Piquet's characters expresses a deeper conflict within American geopolitical culture between unilateralism and multilateralism. Reigart's loyalty is ultimately to his immediate military family and he places that before his position in a multilateral chain of command. Though his actions 'break the rules', he retires with 'honor and respect'. This heroic Jacksonian story line is the same one followed by the Bush administration in its aggressive push for war against Saddam Hussein's Iraq. Though it tries to follow the protocols of diplomacy and seek the approval of the United Nations Security Council for its actions, it ultimately decides that it must follow the 'higher good' of 'defending the nation against 'immanent danger' and launch a pre-emptive war without UN approval. Reigart's conflict with Piquet could easily be that between the United States and France over the invasion of Iraq. One is willing to defy world authority to pursue a noble end. Righteous conviction trumps diplomatic niceties. In Bush's words 'America will never seek a permission slip to defend the security of our country'.[57] Better to be aggressively proactive than to be forever 'watching' and training without doing anything.

This is a militarist tendency and it has been given free expression in the military strategy of the Bush administration. Under Donald Rumsfeld, the role and ranks of US Special Forces has expanded dramatically. Representing the most 'gung ho' element of the US military, Special Forces are specifically trained for secretive missions like high-risk rescues or assaults on high-level enemy targets. Some forces, like the Green Berets, work with indigenous fighters and are generally deployed as part of 'hearts and minds' campaigns. They tend to be embedded in local contexts and act in a situational manner. Other forces are involved in direct-strike actions and have no interest in local linkages.[58] Under Rumsfeld, these units advised by the Israeli military,[59] have expanded considerably with the formation of 'hunter-killer teams' designed to undertake 'manhunts'. They represent the purest form of Jacksonian militarism: a reduction of the world to aggressive warfare and the unleashing of pleasurably righteous violence on perceived enemies in the name of preemptive action.

Third, *Behind Enemy Lines* celebrates this release of righteous violence on America's enemies. Its final shoot-out scene provides the pleasure of seeing the bad guys get blown away by an airborne 'US cavalry' and its high tech weaponry. The message is also pleasurable. Those that track and hunt down vulnerable and isolated Americans will ultimately have to face the wrath of the American military which destroys with overwhelming force. There is no use in negotiation. America has implacable enemies and one should not be fooled by their media pronouncements and public relations. These forces need to be hunted down and brought to justice. All this chimes with the sentiment of George W. Bush's preemptive war against international terrorism. Where *Behind Enemy Lines* departs somewhat from the post-9/11 geopolitics of the Bush administration is its closing legitimation of the idea of 'war criminals brought to account in an international criminal court'. Established in the Hague in March 2003, the International Criminal Court was long resisted by the US government. The Bush administration has refused to join and has, at the same time, pushed to cut the funding of the International Criminal Tribunal on the Former Yugoslavia.[60] Yet the Bush administration articulates the well-worn notion in American geopolitical culture that the interests of the United States and those of the world are the same. In Bush's words, 'the cause we serve is right, because it is the cause of all mankind'.[61]

Movies are, as noted at the outset, ultimately commercial products designed to make money. In this respect *Behind Enemy Lines* has been a successful product which has made lots of profit for News Corp and its producers. To what extent it boosted recruitment to the US Navy, in the same way as *Top Gun* reportedly did, is uncertain, given the overdetermining impact of 9/11. What is clear is that those inspired to join the US military as a result of movies like *Behind Enemy Lines* or those emotionally engaged by its Jacksonian story line and militaristic messages are currently negotiating a world that is a great deal more complex than that presented by the movie and their 'bring 'em on' leaders.[62] Entertainment conglomerates in alliance with military institutions can project geopolitics onto a cinematic screen as heroic masculine war fighting in exotic locales. But when the lights come up, the world's incorrigible complexity remains and the frustrations of geopolitics persist.

ACKNOWLEDGEMENTS

I want to thank the anonymous reviewers of this paper for taking the time to provide detailed and constructive comments. Thanks also to the participants in the National Science Foundation 'Southern Tier States' workshop in Portoroz, Slovenia, June 2004 for their feedback when this paper was presented.

NOTES

1. The original release date of the movie was January 2002. In a tie-in with the movie, featuring ads and excerpts for its production, the Discovery Channel re-broadcast a BBC documentary about O'Grady two days before its release with the new title 'Behind Enemy Lines: The Scott O'Grady Story'. This action was the subject of a law suit by O'Grady in 2003. The budget figures are from 'Producers Commentary' on the DVD release of *Behind Enemy Lines* and from the web sites <http://www.worldwideboxoffice.com> and <http://www.digitallyobsessed.com/showreviews>.

2. In his DVD commentary, Davis notes that the film would not have been made if it were not for these retired military pilots who believed that the film must be made and encouraged him to 'keep at it' until Department of Defense cooperation was secured.

3. The 'adolescent male' demographic is actually more complex than the label suggests. It comprises an audience of youth and adults, males and females, who enjoy stories of masculine crisis, recovery and redemptive re-masculinisation. The imperiled phallocentric order is restored. The film's gendered narrative arc embodies the culture of American masculinity crisis/renewal described by James William Gibson, *Warrior Dreams: Violence and Manhood in Post-Vietnam America* (New York: Hill and Wang 1994) and others.

4. The initial script was largely scrapped because Moore thought it was 'garbage': 'all the bad guys were Chetniks with long fur coats and curved knives. I got rid of all that and made them very complex people with cell phones.' John Moore, 'Behind *Behind Enemy Lines*: Press Conference', <http://actionadventure.about.com/library;weekly/2001;aa113001a.htm>, 2001.

5. Film plot summary, available at the film's official web site <http://www.behindenemylines-movie.com> ,accessed 28 October 2004.

6. The movie is full of commercial features and product placements, from a lingering shot of Hackman's character smoking a cigarette to a bottle of Coca-Cola as a cure for thirst to Northrop Grumman satellite and weapons systems. The most ostentatious product placement of all is the use of News Corp's Sky News and one of its senior reporters as the vehicle for news in the movie, even to the extent of having Sky News broadcasting in America.

7. On the military–industrial–entertainment complex see James Der Derian, *Virtuous War* (Boulder, CO: Westview 2001). David Campbell notes that the story of O'Grady 'behind enemy lines' is reminiscent of captive narratives in American history (colonists in the hands of Indians, for example). See David Campbell, 'Contraditions of a Lone Superpower', in David Slater and Peter Taylor (eds), *The American Century: Consensus and Coercion in the Projection of American Power* (Oxford: Blackwell 1999) pp.222–40.

8. Walter Russell Mead, *Special Providence: American Foreign Policy and How it Changed the World* (New York: Routledge, 2002). I should point out that my terminology is different from that used by Mead. What he terms 'the American foreign policy tradition' I term American geopolitical culture and what he terms as 'paradigms' and 'schools' I term 'traditions'. It is outlined and defended in Gearóid Ó Tuathail, 'Geopolitical Structures and Geopolitical Cultures: Towards Conceptual Clarity in the Critical Study of Geopolitics', in Lasha Tchantouridze (ed.), *Geopolitical Perspectives on World Politics* (Winnipeg: Centre for Defence and Security Studies 2003). US foreign policy is not made up exclusively of any one of these traditions and I do not mean to imply that by focusing in this essay on the Jacksonian story line.

9. To view the image go to <http://www.behindenemylinesmovie.com/wallpapers/800.html>, accessed 29 October 2004.

10. It has also elements in common with post-Cold War remasculinising fables like Independence Day. See Cynthia Weber, *International Relations Theory: A Critical Introduction* (London: Routledge 2001).

11. Susan Jeffords, *The Remasculinization of America: Gender and the Vietnam War* (Bloomington, IN: Indiana University Press 1989) p.73.

12. Jeffords acknowledges how these films tie into what she calls 'other kinds of social thematics currently in operation in American culture: among others militarism, patriotism, individualism, and recuperation of America's international image' (ibid. p.135).

13. Ibid. p.5. On American 'patriot' discourse see Carolyn Gallaher, *On the Fault Line: Race, Class and the American Patriot Movement* (Lanham, MD: Rowman and Littlefield 2003) and Mathew Sparke, 'Outsides Inside Patriotism: The Oklahoma Bombing and the Displacement of Heartland Geopolitics',

in Simon Dalby and Gearóid Ó Tuathail (eds), *Rethinking Geopolitics* (London: Routledge 1998) pp.198–223.

14. This reference to 'women' is purely gestural. There are no women characters in *Behind Enemy Lines*. There are only two brief lines spoken by women throughout the whole movie, one by a member of the USS *Carl Vinson* crew and another by a Bosniak.

15. The male-only nature of this tradition has changed over the last two decades as its violent misogynist masculinity has been exposed, and previously all-male military institutions have been forced to admit women recruits.

16. The US Department of Defense denies it produces 'propaganda films' and will point out that private film producers, rather than the department, make films like *Behind Enemy Lines*. Furthermore, film makers are free to produce any kind of film they so wish. However, few films secure the cooperation of the US Department of Defense. The DoD will not cooperate with any film makers without approval of their script beforehand. Obtaining approval is very difficult and requires film makers to hire influential retired military officials and, if possible, also have powerful political influence. Once approval is obtained, film makers are required to work with scripts acceptable to the military. Particular scenes and images are vetoed if they do not meet with the approval of the military. In *Behind Enemy Lines* a scene where Reigart picks up the phone and calls Sky News to leak the story of the downed pilot was vetoed as it showed a US Navy commander deliberately breaking the 'chain of command'. Instead, Reigart is shown picking up a Sky News reporter's business card. The next scene shows the breaking of the story on Sky News. Yet, interestingly, the American Navy were fine with the overall story which celebrates American unilateralism and the renegade actions of a commander. For a discussion of the process of cooperating with the US military in the making of a film, see the writer's commentary on the DVD of *Tears of the Sun* (Sony Pictures, 2003).

17. David Halberstam, *War in a Time of Peace* (New York: Scribner 2001) p.46.

18. Colin Powell, *My American Journey* (New York: Ballantine 1995) p.559.

19. Halberstam (note 17) pp.38–9.

20. Powell, *My American Journey* (note 18) p.559.

21. 'Now, the war in Bosnia-Herzegovina and Croatia is a complex, convoluted conflict that grows out of age-old animosities. The blood of innocents is being spilled over century-old feuds. The lines between enemies and even friends are jumbled and fragmented. Let no one think there is an easy or a simple solution to this tragedy. The violence will not end overnight, whatever pressure and means the international community brings to bear. Blood feuds are very difficult to resolve.' George H.W. Bush, 'Remarks and an Exchange with Reporters on Departure from Colorado Springs, 6 August 1992', *Public Papers of the Presidents: George Bush* (US Government Printing Office 1992).

22. David Rieff, *Slaughterhouse: Bosnia and the Failure of the West* (New York: Simon and Schuster 1995) p.160 and Roy Gutman, *A Witness to Genocide* (New York: Macmillan 1993) p.13.

23. This phrase and argument was first used by British Foreign Secretary Douglas Hurd and subsequently ad nauseam by Prime Minister Major in defence of the one-sided arms embargo. See Brendan Simms, *Unfinest Hour: Britain and the Destruction of Bosnia* (London: Penguin 2001) p.80.

24. The *New York Times* editorialised: 'The war in Bosnia is not a fair fight. It is not war. It is slaughter. Yet American officials, despairing of a way to end the war, continue to dither. They remain oddly unwilling to acknowledge that even if collective military intervention cannot readily compel a cease-fire, it can at least slow the slaughter.' 'Editorial: At Least Slow the Slaughter', *New York Times*, 4 October 1992, p.D16. Colin Powell records in his memoirs how, when he read this editorial, he 'exploded' and 'dashed off a blistering response'. It was published as an op-ed piece four days later. Powell, *My American Journey* (note 18) pp.558–9.

25. Michael Gordon, 'Bush Backs a Ban on Combat Flights into Bosnian Airspace', *New York Times*, 2 October 1992, p.A1.

26. John Goshko, 'Bush Urges Flight Ban over Bosnia', *Washington Post*, 3 October 1992, p.A1.

27. William Schmidt, 'Britain Wary of Using US Air Power in Bosnia', *New York Times*, 5 October 1992, p.A13.

28. Colin Powell,'Why Generals Get Nervous', *New York Times*, 8 October 1992, p.A35. This reasoning was one of the biggest military blunders Powell made during his time as chair of the Joint Chiefs of Staff. The war, of course, was ended three years later when limited military force was used by the United States and its allies.

29. Paul Lewis,'UN Bans Flights in Bosnia but is Silent on Enforcement', *New York Times*, 10 October 1992, p.A3.

30. Elaine Sciolino, 'US May Seek the Use of Force to Stop Serb Flights over Bosnia', *New York Times*, 4 December 1992, p.A1.

31. Media reports at the time of the ban stated that there was only one confirmed combat violation of the ban. However, other media reports indicated that the Serbs used aircraft to bomb Sarajevo as well as towns and small villages in eastern Bosnia like Osatica (south of Srebrenica on the Serbian border). Julia Preston, 'UN Ready to Enforce Flight Ban in Bosnia', *Chicago Sun Times*, 19 March 1993, p.40; Paul Lewis, 'UN Moving to Toughen Yugoslav Flight Ban', *New York Times*, 19 March 1993, p.A11.

32. Paul Lewis, 'UN Approves Plan to Enforce Bosnia Flight Ban', *New York Times*, 1 April 1993, p.A12.

33. Alan Riding, 'NATO Agrees to Enforce Flight Ban over Bosnia Ordered by UN', *New York Times*, 3 April 1993, p.A5.

34. John Pomfret, 'NATO Jets Bomb Serb Airfield', *New York Times*, 22 November 1994, p.A1.

35. Mary Pat Kelly, *'Good to Go': The Rescue of Capt. Scott O'Grady, USAF, from Bosnia* (Annapolis, MD: Naval Institute Press 1996) p.11.

36. Michael Ryan gives the figure of twenty SAMs being fired at NATO aircraft from November. Most were portable shoulder propelled SA-7 ('man-pads') that were generally not very threatening. A few were SA-6s which were launched from vehicles and could deliver up to three radar-guided surface-to-air missiles with much bigger payloads. The rules of engagement stated that NATO aircraft could go after a SAM site immediately and only if it was a 'smoking gun.' They were not allowed to take out all possible SAM sites and radar facilities.

37. Kelly (note 35) p.15.

38. Ibid. p.16.

39. Though O'Grady did very little beyond stay out of sight, publishers still managed to generate two books about his adventure: Scott O'Grady, *Return with Honor* (New York: Doubleday 1995); and a book for children: Scott O'Grady, *Basher Five-Two* (New York: Yearling 1997). O'Grady left the active military after his adventure and moved to Plano, Texas where he began bible study at Dallas Theological Seminary. He established a career as a motivational speaker for Christian audiences, commanding up to $15,000 a speech. His speeches are paeans to God, family and country. For a video excerpt see <http://www.washingtonspeakers.com/speakers/Speaker.cfm?SpeakerID=1654>, accessed 10 November 2004. O'Grady was also active in the second presidential campaign of George W. Bush, condemning John Kerry's actions during the Vietnam War as 'treason'. See Minnesota Public Radio, 'O'Grady: Kerry's Actions after Vietnam Constituted Treason', available at <http://news.minnesota.publicradio.org/features/2004/08/13_ap_ogrady/>, accessed 10 November 2004.

40. Kelly (note 35) p.31.

41. Ibid. pp.170–71.

42. Ibid. p.294.

43. In a press conference at the release of the film, director Moore denied any connection to the O'Grady story beyond the fact that it is about an American aviator shot down in Bosnia. 'We certainly didn't approach Captain O'Grady to make his life story. I never met Captain O'Grady and I never researched material on him.' One significant difference between the film and O'Grady's story is the absence of religious discourse in the film. O'Grady interpreted his rescue in evangelical Christian terms: 'God veiled me from them'; 'my precious child, I would never leave you'; 'if it wasn't for God's love and my love of God, I wouldn't have got through it. He's the one that delivered me here'; O'Grady (note 39) pp.89, 105, 186. So also did some members of the military: 'The Lord was watching over Captain O'Grady, and eventually he was watching over the Marines and sailors when we went into a combat situation'; Kelly (note 35) pp.319, 325–6. Kelly's account of the rescue begins with the biblical quotation: 'Who among you, if he has one hundred sheep, and loses one, will not leave the ninety-nine to follow the lost one? And when he finds it, he puts it over his shoulder and carries it home in jubilation' (Luke 15:4–5). In August 2002, O'Grady filed a law suit against Fox Searchlight Pictures claiming that it failed to secure the rights to his story when it made *Behind Enemy Lines*. He also sued the Discovery Channel for using a documentary about him as a tie-in to the movie.

44. For example, Colonel Chuck Wald, commander of the Aviano air base in Italy, stated that 'I'm afraid they'll fight in Bosnia until they end up fighting with spoons if they have to. Any resolution in Bosnia must come from the combatants'; Kelly (note 35 p.48). Scott O'Grady's knowledge of the Bosnian war came from US Air Force intelligence in Aviano: 'there were no friendly forces in Bosnia, no safe areas, no one you could count on. This was not World War II, when front lines meant something and the French Resistance harbored many an American pilot. This was modern warfare, where it seemed that the only thing united a crazy quilt of factions was a shared distrust of the U.N. peacekeepers and the

NATO airpower behind them'; O'Grady (note 39) p.43. This is the only comment on the Bosnian war in his memoir.

45. David Campbell reads the willingness of the US military to deploy a substantial force to save one man, in contrast to its reluctance to deploy any forces to respond to the deaths of more than 200,000 Bosnians, as exemplary of the contradictions of the United States' engagement with the post-Cold War world. The US military, like most militaries, is a bureaucracy with a strong line between insiders and outsiders. Given the role of Christianity in also determining that line for many, identifying with Muslims may have been difficult.

46. In his Bosnia memoir Richard Holbrooke recalls that Leighton Smith personally did not agree with NATO bombing in Bosnia and told him, echoing James Baker, that he did not have a 'dog in this fight'. In Holbrooke's view 'Smith was edging into an area of political judgments that should have been reserved for civilian leaders'; Richard Holbrooke, *To End a War* (New York: Random House 1998) p.118.

47. Carl Bildt, *Peace Journey: The Struggle for Peace in Bosnia* (London: Weidenfeld and Nicholson 1999).

48. Holbrooke (note 46) pp 335–7.

49. One of the interesting features of the US deployment in Bosnia is how ending it became a cause célèbre in right-wing military circles in the United States. For years, Texas Senator Kay Bailey Hutchinson spoke out against the deployment in the US Senate and sponsored legislation to withdraw the troops. George W. Bush also called for US troops to be withdrawn from Bosnia when he was running for president. Why this was the case is puzzling but it appears that many Capitol Hill Republicans never accepted that Bosnia or later Kosovo was worth US involvement. The wars there were seen as Clinton/Albright 'peacekeeping' wars. That they were on behalf of Muslims may account for the opposition of some. The Balkan deployments supposedly symbolised a 'readiness' problem in the US military. It was needed for war fighting in 'defence of the nation', not foreign policy 'social work' like peacekeeping. Furthermore, those making these arguments held that there was no US national interest in the Balkans because there was no oil there or anywhere nearby. The views of US military commanders after the Bosnian war contradict these positions. One report found unanimous agreement that US peace operations like Bosnia strengthened US leadership and did not comprise war-fighting ability. In fact, the highest reenlistment in the army was from soldiers serving in Bosnia. See Edith Wilkie and Beth DeGrasse, *A Force for Peace: US Commanders' Views of the Military's Role in Peace Operations, A Project of the Peace Through Law Education Fund*, 1999, available at <http://www.ptlef.org/reports.htm>, accessed 10 November 2004.

50. Sherri Sylvester, 'Hackman, Wilson hit the (Ship) Deck for "Enemy Lines"', *CNN Showbiz Today Reports*, 28 November 2001.

51. 'Producer Commentary' on the DVD of *Behind Enemy Lines* and 'Production Notes' on the official film website <http://www.behindenemylinesmove.com> See notes 13 and 14.

52. On the importance of gear in the new militarism see Thomas de Zengotita, 'The Romance of Empire and the Politics of Self-Love', *Harpers Monthly* (2003) pp.31–9.

53. Michael Mann, *Incoherent Empire* (London: Verso 2003).

54. For a critique of this threat inflation from a military strategic point of view see Jeffrey Record, 'Bounding the Global War on Terrorism', Strategic Studies Institute, US Army War College, Carlisle, Pennsylvania, available at <http://www.carlisle.army.mil/ssi/pubs/2003/bounding/bounding.htm> ,accessed 21 January 2004.

55. Kevin Baker, 'We're in the Army Now: The G.O.P.'s Plan to Militarize Our Culture', *Harper's Magazine* (2003) pp.35-46.

56. Mann (note 53) pp.7–8.

57. George W. Bush, 'State of the Union Address', 20 January 2004, available at <http://www.whitehouse.gov>, accessed 21 January 2004. During the 2004 presidential campaign George W. Bush charged that John Kerry would subject the use of US force to a 'global test'. The charge was a gross distortion of Kerry's position but Bush used it and explicit xenophobic demagoguery to establish the perception that he was the authentic organic American and Kerry an aloof Europeanised American (who spoke French). See Sandra Maler, ''French' Becomes Campaign Slur', *The Seattle Times*, 27 October 2004, p.A5; and Clarence Paige, 'The Attack on Kerry's "Global Test"', *Baltimore Sun*, 7 October 2004, p.23A.

58. Gregory Vistica, 'Military Split on How to Use Special Forces in Terror War', *Washington Post*, 5 January 2004, pp.A1, A9.

59. Seymour Hersh, 'Moving Targets', *New Yorker*, 15 December 2003, pp.48–55. Hersh reports that one of the key planners of the Special Forces offensive is Lieutenant General William Boykin.

Boykin shot to prominence in 2003 when his remarks comparing the Muslim world to Satan became public. In Sunday morning talks in uniform to church groups he reportedly cast the war on terrorism as a religious war and hailed President Bush as 'a man who prays in the Oval Office' and was 'not elected' but 'appointed by God'. The Muslim world hates America 'because we are a nation of believers'. Despite calls for his dismissal, Boykin was not fired. William Arkin, 'Commentary: The Pentagon Unleashes a Holy Warrior', *Los Angeles Times*, 16 October 2003, p.B17.

60. Keith Richburg, 'International War Crimes Court Is Inaugurated, but Without U.S.', *Washington Post*, 11 March 2003, p.A18. *New York Times*, 'Editorial: Faster Justice for the Balkans', *New York Times*, 28 November 2003, p.A30.

61. Bush, 'State of the Union Address' (note 57).

62. Gearóid Ó Tuathail, '"Just Out Looking for a Fight": American Affect and the Invasion of Iraq', *Antipode: A Radical Journal of Geography* 35/5 (2003).

Staging the Border: National Identity and the Critical Geopolitics of West German Film

INGA SCHARF

Department of English and Media Studies, Nottingham Trent University, Nottingham, UK

INTRODUCTION

This essay examines what impact the geopolitical changes occurring in the aftermath of the Second World War had on West German national identity constructions at the time. 'Geopolitics' is understood as 'practices and representations of territorial strategies',[1] which function as a means of forming and maintaining identities in modern societies. In relation to this broadly cultural conceptualisation of 'geopolitics' as a complex web of representational and identity constitutive practices, (national) cinema becomes a prime research target. As such, it can be regarded as a specifically symbolic system which serves as an important means of cultural (re-)production including, amongst other things, elements of the geopolitical realm.

More precisely, the investigation sets out to analyse films of the *New German Cinema* (NGC)[2] – a West German cinema movement that critically engages with traditional and problematic ways of German self-narration – which, as demonstrated below, can be seen as being valuable popular sources to elaborate on West German national identity issues taking a distinctly geopolitical approach. Thus the concept of 'geopolitics' is discussed in relation to or, more accurately, *through* an examination of the narrative strategies of these films since, in the light of the above definition, the latter are crucially seen as geopolitical narratives in themselves. However, looking more closely at the films' representations of 'homelessness', 'border' and 'not belonging', it needs to be stressed that the NGC's 'geopolitics' is an inherently critical one. To illustrate this point further, Ó Tuathail's concept of 'critical geopolitics' may serve as a useful trajectory.[3]

'CRITICAL GEOPOLITICS' AND THE NEW GERMAN CINEMA

In his introduction to *Rethinking Geopolitics* Ó Tuathail refers to five aspects of 'critical geopolitics' which can be summarised as follows. First, he describes the traditional understanding of geopolitics, which focuses on the actions of '"wise men" of statecraft', as being too narrow, and instead calls for an opening up of geopolitics in theory and practice. Here, his argument resonates with Sharp's suggestion that 'in critical studies, the production of discursive practices of geopolitics should be investigated in locations that may lie outside of the formal arena of state'.[4] Following Bhabha, Ó Tuathail further argues that 'counter-narratives of the nation are forms of critical geopolitics'.[5] Second, he states that 'critical geopolitics' are especially concerned with 'the boundary drawing practices and performances that characterize the everyday life of states'.[6] Third, Ó Tuathail advocates a three-dimensional structure of geopolitics which according to a critical perspective consists of 'practical', 'formal' and 'popular' strands.[7] Fourth, he describes 'critical geopolitics' as 'situated knowledge' which, for example, acknowledges the male-dominated character of traditional geopolitical theories and practices.[8] The latter approach is further developed in Gilmartin and Kofman's concept of 'feminist geopolitics' within which the authors draw attention to the gendered character of geopolitics itself.[9] Finally, Ó Tuathail asserts that 'critical geopolitics' also needs to be aware of its disciplinary context and historical development.[10]

Relating these reflections on the concept of 'critical geopolitics' to the filmic as well as socio-political practices of the NGC, it can be said that all five points raised above resonate with respective elements of this cinematic movement. First, the NGC provides a critique in the form of national counter-narratives of the post-war West German society, about which Elsaesser says: 'What made this critique original was that it was not formulated in the terms of a critique, but as a playing out ... It brought into view a stark antithesis around which the first German postwar generation conceived many of its personal narratives'.[11] Second, the NGC offers some very interesting insights into the construction and representation of inter-personal and inter-national 'borders', as this essay demonstrates using three NGC productions as typical case studies. Third, the different layers of geopolitics specified above are present in the chosen NGC films, even though the 'formal' and 'practical geopolitics' of the Cold War are largely assumed as a background in front of which the three cultural narratives unfold. The German–German 'border' is thus depicted as a manifestation of (relative to the films' present) earlier geopolitical theories and practices whose traditional cultural meaning is fundamentally challenged by the use of popular media such as a newspapers, photographs and art, and a pirate radio station amongst other personal strategies. Fourth, the investigation of these films shows them to be gender sensitive in the sense that female and male geopolitical strategies are portrayed as taking substantially different forms. Finally, tracing the history of the concept of geopolitics from the late nineteenth century to the present day, it becomes apparent that the time period of the films (dating from 1977 to 1982), which in themselves are expressions of 'popular geopolitics', can simultaneously be described as a geopolitical 'zero hour' with regard to academic work or 'formal geopolitics'.[12] Between its first highpoint at the beginning of the twentieth century and its second following the end of the Cold War in 1989, the term became discursively unavailable after its appropriation by the Nazis, and was only slowly rediscovered by French geographers in the late 1970s and by Anglo-American academics in the 1980s. Hence, by turning to a phase which has not been studied widely in geopolitical terms, as one can argue at least for the German case,[13] this investigation closes an analytical gap, following the geopolitical conviction of the 'central place of visualization',[14] according to which filmic sources are seen as providing 'exemplary geopolitical image(s)'.[15]

To elaborate further on the specific relationship between geopolitics and film in addition to the introductory remarks following Gilmartin and Kofman, the work of Dodds and Ó Tuathail (1994) has to be considered.[16] Dodds starts from within Ó Tuathail's 'critical geopolitics' model by firmly positioning film (and other media) in the sphere of the 'popular'. Under the heading 'films and geopolitical visions' he then discusses their relationship as one in which film is seen as a powerful tool for the spread of geopolitical 'propaganda'.[17] A heightened version of this tendency to assume a rather affirmative and uncomplicated relationship between the nation-state and (its) national cinema can also be found in Ó Tuathail's account, where Griffith's *The Birth of a Nation* (1915), Eisenstein's *October* (1927) and Riefenstahl's *Triumph des Willens* (Triumph of the Will, 1935) are considered 'remarkable works of geopolitical propaganda'.[18] Moreover, the author writes, 'the shooting camera was from the beginning a weapon of war', a quotation which illustrates the dominant reading of cinematic productions as being highly ideological. Interestingly, such a reading is not adapted in the case of, for example, Bell's cartoons,[20] which are seen as 'visual critiques'[21] and '"geopolitical texts", which illuminate or even subvert particular political practices'.[22] Yet it is just this critical potential, such as that granted to these comic strips, that essentially informs the NGC productions analysed below. Accordingly, this essay might be seen as a contribution to a new reading of films as 'critical geopolitics' in a popular format.

There are very few West German films between 1961 (the year of the 'cementation' of the German–German border in the form of the Berlin Wall) and 1989 (the year of the German reunification) that deal with depictions of the border between East and West Germany. On the side of the relevant literature, this topical neglect is met by a virtual non-existence of book chapters or subsections on 'border' issues. Thus, if one looks at the work of leading authors in the field such as Elsaesser, Corrigan, Hake, Pflaum and Prinzler, Sandford, Franklin, Kaes, and Phillips,[23] one notices that their writings largely concentrate on the study of a set of films, which are regarded as characteristic of the NGC, or a number of filmmakers (usually exclusively male and with Fassbinder, Herzog and Wenders at their core), who again are seen as representative for the movement as a whole. If 'border' issues are mentioned at all they are usually referred to in passing, as for example in Elsaesser's historical account: 'In the films that deal with borders and border-crossings, with journeys, exiles and exotic adventures, with America in Europe, or with a divided Germany, historical traumas are never far away'.[24]

This quotation, which in the book leads on to a discussion around symptomatic individuals and their relationship to the post-war West German society, strikingly illustrates how the 'by-passing' of 'border' issues is articulated with deep-seated 'historical traumas' or, in other words, the overpowering past one has not yet come to terms with. Hence similar to the history of 'geopolitics' whose Nazi appropriation[25] prevented the concept's further development or even use for almost 30 years, most NGC films and the literature surrounding them are still largely dominated by Holocaust-related issues. As a result, the narration of the German–German border or the Berlin Wall as a form of 'critical geopolitics' problematising the complex past–present relationship between and among post-war West Germany, its Nazi predecessor and the Allied Forces is not given much scope. Pflaum and Prinzler speculate that the relative lack of 'border' representations might be due to the Cold War situation within which neither East nor West Germany dared to upset either side of the Iron Curtain.[26] Ironically, the authors exacerbate this absence by only providing rather brief plot summaries of a couple of films alongside a few short references to others productions.

In this essay, this thematic absence is addressed by examining three symptomatic, yet relatively neglected 'border narratives'. Made in the late 1970s and early 1980s, they provide a complex treatment of the topic of 'separation', 'border', and above all the Berlin Wall, placing the German–German division at the core of their story: Schilling's *Der Willi Busch Report* (The Willi Busch Report, 1979), Sander's *Die allseitig reduzierte Persönlichkeit – Redupers* (Redupers, 1977),[27] and Schneider's *Der Mann auf der Mauer* (The Man on the Wall, 1982).[28] The selection process was guided by the aim to choose films whose narrative strategies identify the latter as 'visual critiques'[29] and '"geopolitical texts", which illuminate or even subvert particular political practices'.[30]

In terms of these films' and their filmmakers' national cinematic context, both can be associated with the NGC which was producing counter-narratives to the traditional (and problematic) forms of German self-narration as perpetuated in, for example, 'Heimat' film, a genre which, with regards to production and reception figures, dominated the 1950s' cinematic scene in West Germany.[31] (Well known examples are *Schwarzwaldmädel* (Black Forest Girl, Deppe 1950) and *Grün ist die Heide* (Green is the Heath, Deppe 1951). Both of these films, like several other 'Heimat' productions, are remakes; in this case they refer back to productions made in 1933 and 1932[32] – a fact which indicates a rather backward-looking tendency of

'Heimat' films and shows their problematic relationship with the Nazi past. In narrative terms, productions of this genre perpetuate an unproblematic sense of place centring on the concept of 'Heimat', a romantic ideal central to German national identity constructions for centuries.[33] In short, 'Heimat' combines notions of 'embeddedness', 'wholeness', 'purity', 'happiness', 'fulfilment' and 'peace' and thus dreams up a world desired for but never achieved or achievable. Its predominant representation (not only) in 'Heimat' films takes the form of belonging to a rural idyll usually located within an archetypal German landscape: typically the Black Forest, the Lüneburg Heath or the Alps. Morley and Robins state that the idea of 'Heimat' is 'drenched in the longing for wholeness, unity, integrity'.[34] In view of that, the subsequent investigation of issues such as 'homelessness', 'division' and 'not belonging' in the form of a close analysis of the three NGC 'border' narratives is substantially prefigured by a powerful national/cinematic 'Heimat' discourse which marks out the NGC productions as being programmatically anti-'Heimat'.

Founded in 1962, the emergence of the NGC coincided with the decline of the 'Heimat' film genre which, amongst other developments, brought about a deep crisis of the West German film industry. Furthermore, it directly followed the erection of the Berlin Wall in 1961 – an incident which further cemented the pre-existing socio-political split between East and West Germany. Thus the foundation of the NGC has to be understood as a reaction to this cinematic and geopolitical crisis in which a group of young West German filmmakers who had been successful at international festivals claimed the death of, in their opinion, an outdated form of German cinema and announced the birth of a new one (thus the title *New German Cinema*). They expressed their cinematic aims and intentions most notably in the so-called Oberhausen Manifesto in which they recognised the powerful role of narratives in people's lives and thus in society but did not agree with the '(his)stories' that were told by others about Germany (see the dispute about the American series *Holocaust* discussed in Eric Santner's 1990 account)[35] or even by German filmmakers of the immediate post-war years whose productions they despised as 'Opas Kino' (grandpa's cinema). Being part of this cinematic movement, the chosen NGC productions are, in contrast to 'Heimat' films, less part of a mass form of popular culture customised for entertainment purposes than a critical and (nationally as well as internationally) audible, though not dominant, voice geared at socio-political change. Analysing the films in the order below will lead the investigation from being gradually closer to, then 'at', to finally 'on' the Berlin Wall/'border', and thus to the evaluation of its significance for the post-war German context.

Post-war West Germany can, then, be conceptualised as a complex site of struggle over meaning, which is affected by various forms of geopolitics from the 'national inside' as well as the 'international outside': the German–German border being the major, yet not the only, point of interference regarding these two geopolitical spheres for which the films indicate a narrative correspondence. In this vein, the analysis sets out to cast light on the particular set of meanings that has been ascribed to the geopolitical construction called Germany, a place that, up until the end of the Second World War, was imagined primarily in terms of 'Heimat'. However, as Morley and Robins point out, post-war Germany became 'represented by its division, above all by the Berlin Wall'.[36] With the latter as its national symbol, Germany became firmly positioned at the centre of the major post-war geopolitical conflict known as the 'Cold War' and was placed in the middle of a national identity crisis around 'belonging'.

PERIPHERY TAKES CENTRE STAGE

Schilling's *Der Willi Busch Report* (The Willi Busch Report, 1979) is an ironic humoresque about a small town close to the German-German border (Friedheim in Hessen), the role of the local newspaper (*Werra-Post*) and the never-ending efforts of a publishing family to sustain the paper and with it themselves (Adelheid and Wilhelm Busch – brother and sister). Right at the start a skilful collage with voiceover is presented to establish the film's geopolitical context and link world politics and individual existence as follows:[37]

> *The global level*: (documentary material in black and white)
>
> There are Churchill, Roosevelt and Stalin in Yalta in February 1945. During the conference of Yalta they discussed what to do with Germany after its defeat. They did not reach an agreement, but all three were heavily convinced that Germany would have to be split into various parts in order to provide for the future.
>
> *The European level*: (a European map is cut into two parts with a pair of scissors)
>
> Very soon after the end of the war a strange line grew dividing the middle of Europe.
>
> *The national level*: (the actual feature film footage starts with the portrayal of the German-German border)
>
> A border that became increasingly impenetrable by the day shoved certain villages and cities of a country from its centre to its remotest periphery.
>
> *The personal level*: (the camera circles around a male figure dosing in the grass)

> And that is Wilhelm Busch.
>
> ...
>
> *The economic level*: (the camera continues to be on the protagonist)
>
> With the birth of little Willi a slow decline of the Werra-Post began. Although Herman Busch (Willi's father) managed to manoeuvre his paper with great effort through National Socialism and the war, he soon had to realise that this new border, which was literally on his doorstep, had cut him off from two thirds of his market.
>
> *The local level*:
>
> As a result, the once lively Friedheim turned into an eventless provincial town.

This initial commentary of the film shows how distant geopolitical decisions (conference in Yalta) between three men (Churchill, Stalin and Roosevelt) can shape the character of a whole continent (Europe), which again affects the condition of the nation-states belonging to it and (especially in the case of Germany) their inner structure down to the level of towns (Friedheim) and finally the everyday life of their inhabitants (Willi Busch and his *Werra-Post*). In this respect the filmmaker proposes a direct line between Yalta and Friedheim, Churchill/Stalin/Roosevelt and Willi Busch, the conference of Yalta and the economic position of the *Werra-Post*, 'practical' and 'popular geopolitics'. Thus 'place' is conceptualised as the effect and representation of a complex network of power relations (however emphasising the primacy of 'practical geopolitics'), which interact on different levels: the voiceover/narrator relates distant places, people and events to each other, reveals hidden links and asks the question of centre and periphery within a world portrayed as governed by leading states*men*. In view of the latter these opening comments stress the constructedness of place according to geo-strategic considerations, which in the Friedheim example move a town from the core to the margins of its national frame of reference. In conclusion, the opening sequence of the film suggests that senses of a place are not static but dependent on geopolitical decisions and alliances, which change over time.

The rest of the film then explores the relationship between a decisive change amongst the possible range of available senses of place occurring in Germany in the aftermath of the Second World War – from being centred around 'wholeness' (pre-war) to imaginations which focus on notions of 'border' and 'division' (post-war, Cold War) – and its appropriation by the media/Willi in the case of the *Werra-Post*/Friedheim. Hence the plot elaborates on notions of 'popular geopolitics' by exploring the co-dependency of a place and its media representation in the following way: provided that a local newspaper is only as interesting and significant as its place of production, Willi desperately tries to simultaneously transform Friedheim's sense of place and improve the situation of the *Werra-Post*. His strategy is to (literally) create media events in an eventless provincial town so that he can fill his paper with sensational stories, which will then attract more readers and help to sell it. In other words, in an attempt to counter Friedheim's seemingly sealed geopolitical fate Willi strives for a re-centralisation of the peripheralised place by means of media representation to effectively strengthen the position of the *Werra-Post*.

At first Willi claims that there is a five-year-old girl in Friedheim who is able to predict the future in general and in particular German reunification. Second, when an old friend and former colleague of his unexpectedly dies of a heart attack in his hotel room, Willi blows up the harmless incident into a spy affair. Shortly after, another two deaths follow (a dentist and Willi's gardener) and things become a lot rougher and 'city-like' in Friedheim. Eventually, Friedheim makes it into the headlines of the big national newspapers and thus moves to the centre of the media's interest, becoming renowned as 'the nest of spies'. With its various spectacles the town then increasingly turns into a media site attracting eight buses every day, which carry pilgrims to the child prodigy. As a result, the *Werra-Post* comes out daily instead of three times a week due to a dramatic increase in its print run.

This main part of the film can be seen as problematising and elaborating on the geopolitical framework outlined in the opening sequence. The Second World War and its consequences in the form of Churchill/Stalin/Roosevelt's 'practical geopolitics' have a decentring effect on Friedheim's post-war sense of place, which Willi through 'popular geopolitics' tries to subvert for personal reasons, although with decisive side-effects. However, his resistance and agency in this film are restricted to an initial 'invention of traditions' since the media and tourist industry soon take over and the narratives take on a life of their own. Still, he (in his role as a local journalist) is the catalyst, facilitating the emergence of a new sense of place: During the process of changing Friedheims's image through the creation of spies and fortune-tellers as a means of reviving its inhabitant's interest in their hometown, not only his newspaper profits but also the place as such, which is geographically marginalised, moves into the centre of media attention. This re-centring of Friedheim, however, is only possible as 'border' and 'division' can be seen as indispensable elements of a post-war German sense of place which makes the Friedheim case an example of a shift in self-narration after the division of Germany which can be called 'the (re-)centring of the peripheral'. Thus 'practical geopolitics' on a global level at first had a marginalising effect on Friedheim's sense of place, before notions of 'border' became central to post-war Germany's sense of place, subsequently providing the town with the opportunity to regain significance in the regional and as well as the national imaginary through the production and dissemination of 'border narratives', in other words, 'popular geopolitics'.

Regarding the portrayal of the German–German border, there are constant visual and dialogical references. The mayor of Friedheim, for example, apologises for his town in advance when a new teacher arrives: 'We are a bit remote here.' And Willi's girl-friend shouts full of anger: 'Bloody fence! I can't stand the sight of it any longer!' Quite the opposite, Willi hardly talks about or mentions the border but instead constantly gazes at it through his binoculars. In this respect, the border can be seen as fetishised, with Willi permanently observing what happens on the other side. Considering Laura Mulvey's concept of the 'male gaze'[38] the East (GDR) is thus feminised and objectified[39] – which explains why Willi's girl-friend reacts so furiously at 'the sight of it'.

Hence the 'fence' is not only a cut-off-point, a dividing line or a negating factor but also productive in the sense of arousing one's imagination and curiosity. The border says 'no' and 'yes' at the same time: it says 'no' on a spatial level, peripheralising Friedheim's geopolitical position and de-centring it within a national context. It says 'yes' on a narrational level since it serves as a pretext for reunification dreams and spy stories, which, articulated by the *Werra-Post*, draw on existing Cold War narratives that circulate within post-war Germany. The latter appropriation of discursively available senses of place whose key component or symbolic centre is the border re-centres Friedheim as a representation of typical lived experience within the FRG.

Finally the narrative can be divided into three temporal spatialities, which refer to different senses of place. (1) The pre-war period: it constructs a pre-history for the narrative, which involves the representation of Friedheim as being a central place, full of life and confidently rooted in its geopolitical framework. (2) The immediate post-war period up to the beginning of the film – through spatial changes within Germany, Friedheim is peripheralised, leading to the place's decline in terms of geopolitical significance: it becomes ruralised and eventless and thus effectively lacks narratives to tell about itself. It becomes a non-place. (3). The time period of the film: Willi's 'popular geopolitics' result in Friedheim's repositioning in the centre of the national imaginary – the periphery takes centre stage. The town is increasingly urbanised, eventful and mediatised.

In effect the two processes responsible for shifts in Friedheim's sense of place are 'peripheralising' (the effect of 'practical geopolitics') and 're-centring' (the result of 'popular geopolitics'). Thus besides the 'border' the film offers a second geopolitical/narrative component, which drives the story: the periphery. Overall the above 'border narrative' focuses on issues of 'division' and 'fragmentation', which in terms of its portrayal of a particular part of rural Germany has a considerably de-romanticising, anti-'Heimat' effect. However it could be argued that the girl's reunification forecast refers to rather hidden but nevertheless available and active dreams and desires for wholeness and 'Heimat'. I will return to this point later.

AN ART(IFICIAL) WALL

With Helke Sander's film *Die allseitig reduzierte Persönlichkeit – Redupers*[40] (Redupers, 1977) this analysis of German 'border narratives' moves to Berlin, the city which was divided by the building of the Berlin Wall in August 1961. It is a story about a West Berlin photographer, Edda Chiemnyjewski, who tries to juggle single motherhood, a demanding job and her involvement in a communal arts project. Being a member of a women's group of photographers she combines her rather private and individual artistic production with forms of cooperative political commitment in general and feminism in particular.

Starting off with a drive through the city and alongside the graffiti clad wall, the film first shows the latter as an indispensable component of the protagonist's everyday life experiences in this particular part of West Germany which happens to be situated in the Eastern part of the country – isolated like an island. In this respect the wall is a constant reminder of this artificial and geopolitically rather unique situation: That one is *outside* the West German territory but still *inside* its political realm. This notion of Berlin's 'special status' (an expression often used in West German political rhetoric), which in some sense can be seen as a euphemism for 'non-place', is further elaborated by a short scene in which the postman brings Edda a letter from the United States onto which the sender has written (in English):

> Postmaster: Berlin has a special international status it is neither in West Germany nor in East Germany. Still, I hope you can find West Berlin and deliver the letter. Meanwhile I suggest that you go to see Carter in the White House to find out where Berlin really lays.

Thus in the vein of *Der Willi Busch Report* the film starts off by drawing attention to the influence of global 'practical geopolitics' on the construction of German territory along with its particular senses of place. However, Berlin's 'special status' in geopolitical terms has become normality for the people who live in the city: Whenever she leaves the house, Edda sooner or later encounters a segment of the wall and when she is at home and listens to the radio there are news stories about people trying to overcome the wall (East–West refugees). Hence the symbolism of the Cold War and Germany's division is ubiquitous, a reality which renders the wall virtually invisible.

Berlin's inhabitants have learnt to cope with their unique living situation, which to a certain extent meant to allow the exception to become the rule. Only some writings on the wall articulate a clear sense of unease and anger in relation to their 'means of communication'. Then again these voices are countered by the mélange of radio stations (East German, West German, English, American, French, etc.), which broadcast their programmes to the population of Berlin, making it a place where cultures, countries and opinions meet rather than are kept apart.

In addition to (passively) being an aspect of lived experiences the Berlin Wall becomes aesthetically appropriated and in this way made productive in the communal arts project of Edda and her colleagues, which thus can be regarded as a form of 'popular geopolitics'. Funded by the local government, its aim is to document the city of Berlin (of which the wall is an essential part) using photographic means. During the preparatory stages the women discuss the city's function and image along with its representation in the media in order to develop their own rationale for the project. However, there are conflicting arguments held by different members of the group: Should the photos be socially critical (serving the public: the ethical position) or 'advertising' (pleasing the financial backer: avoiding censorship)? They finally agree on their selection criterion, this being the photo's function as a means of critical intervention in the public domain, whereby the artists consciously risk public censorship and the prohibition of a planned exhibition.

Exemplary for this understanding of art as an alternative public sphere and a 'critical geopolitical' tool is Edda's photo series, which has three sections: (1) The question of what is similar in the East and West. (2) The question of where the wall has holes, and of where is it transparent. (3) The 'newspaper' character of the wall. Thus in opposition to Willi who (mis)uses the media for personal purposes, the female photographers' actions and artistic interventions are guided by strong feelings of responsibility for the wider Berlin public. They ultimately seek to establish a dialogue with their audience – an aim they try to achieve by creating different 'contact zones' to facilitate an exchange of ideas between their work and potential recipients.

Thus art becomes a specific form of communication which aims to break the predominant social silence. First, the artists photograph writings in various forms (especially inscriptions on the wall, such as graffiti), which are available to Berlin's public. Through that they answer these people's call for communicative interaction and take it further by giving it space in their artistic production which again, as the latter are publicly exhibited, addresses the original call. Second, they do a 'border performance' on a vantage point at the wall itself whereby they install a white curtain which, drawn apart, reveals the other side of the wall (East Berlin). By means of this theatrical production they draw attention to the 'performativity'[41] of geopolitical reality (in this case the wall dividing East and West Berlin) and its social constructedness. As such, they question the 'othering' involved in this process that eventually renders East Berlin, which looks very similar to the Western part of the city, different instead of the same.

Between her different artistic appointments Edda is dealing with a range of job and family related tasks such as driving Aunt Gretchen back to the border. The latter is a so-called 'grey panther aunt' ('grey panther' being a euphemism for pensioner[42]) who brings jeans from the West to the East and cheap potatoes from the East to the West. While exchanging these goods between the two sides of Berlin, she constantly crosses the border and thus repeatedly (re)establishes and maintains a link between the Eastern and the Western part of the city. Hence she can be seen as a messenger and communicator between the two cultural spheres which she simultaneously inhabits – a way of life through which she persistently negates and penetrates the wall. Furthermore, Aunt Gretchen is a 'residue' who besides linking geographical spheres also connects temporal zones: She was born in and experienced Germany as a whole (as opposed to Edda who grew up with the division) and thus carries within her the hope of a German reunification in the future.

On a different level this time–space symbolism is repeated through the presentation of Berlin as a building site: there are heaps of rubble everywhere – a sight which bears resemblance with the typical 'Trümmerfilm' (rubble film) setting right after the Second World War. These generic references link this film of the late 1970s – a time of firmly established geopolitical separation of the two Germanys – to productions of the late 1940s – made a generation ago when the division had not yet been completed. However, besides this link to the past, a building site is always also a metaphor for the future: Something new is just about to be realised; the current state is transitionary. Thus, by referring back to the image of a united past the building site and its rubble then project a hopeful future, which might entail reunification. This is reminiscent of the girl's reunification forecast in *Der Willi Busch Report*.

In conclusion, whereas *Der Willi Busch Report*, made by a male director, depicts men as dividing forces (Churchill, Roosevelt, Stalin) or individually-resistant egoists (Willi), *Die allseitig reduzierte Persönlichkeit – Redupers*, made by a female filmmaker, portrays women as uniting forces (Aunt Gretchen) or dialogue-initiating collectivists (Edda and her colleagues). Hence both films reproduce quite traditional gender roles in relation to different societal realms – 'practical geopolitics' and business-driven 'popular geopolitics' as the spheres of men, and creative 'popular geopolitics' and family as the female areas of interest – and can therefore be seen as complementary narratives.[43] In the end, both films stress the productive side of the border/Berlin Wall while at the same time (although their protagonists reveal different motivations) propagating an atmosphere of hope: reunification seems possible.

NO-MAN'S-LAND

Peter Schneider's *Der Mann auf der Mauer* (The Man on the Wall, 1982) is the only film to explicitly mention the Berlin Wall in the title, indicating that notions of 'separation' and 'division' lie at the core of the narrative. More accurately, the position of a particular man *in relation to* this specific type of border and the issues surrounding it seem to be of crucial importance, which reveals that the film engages with issues of 'relationality' and 'positioning'. In many ways the narrative that follows represents a very radical way of making sense of the place called Germany since the relationship between the spatial and the social, which is certainly a crucial issue within 'critical geopolitics', is constructed as a particularly intimate one.

At the beginning of the film a young man in East Berlin (Arnulf) gazes through his binoculars at the other side of the city – a scene reminiscent of Willi's 'border gazing'. Thus again, 'place' is feminised and objectified (see note 7). However, whereas Willi looks East, Arnulf is looking West – signifying that either part of Germany can be similarly identified with female traits in relation to the male gaze from beyond the fence/Berlin wall (a variation of the gender–place relationship which finds its most common expression in the bond between the soldier and the country he fights for). Besides this implicit feminisation of 'place' there is also a reaffirmation on the level of dialogue presented immediately after the 'silent gaze' when Arnulf tells his wife (Andrea) that he wants to go and live on the other side:

> Andrea: You are leaving me.
>
> Arnulf: No, I am leaving this state.
>
> Andrea: That is the same.
>
> (Silence)
>
> Arnulf: (instead of saying goodbye) Always remember, they [the two parts of Germany] will get back together.

Having finally made it to the West (after one and half year's imprisonment for his illegal attempt to leave the country) Arnulf tries to call Andrea, yet, without getting through to her. He says to himself: 'I will re-establish the connection. Don't worry, Andrea, we will get back together.' Thus he repeats the words he used to express his hope for a German reunification to talk about his and Andrea's future which at that point of the film, associates him with West Germany and her with East Germany – two geopolitical parts of an item which have been split apart and then lost contact with each other.

Shortly after, Arnulf meets another women, Victoria, who is West German and has prejudices against East Germans (she calls him arrogantly 'Zoni' – short for 'from the Eastern Zone'). Hence she can be seen as being out of touch with the other side of Germany/herself. They start having an affair while Arnulf's homesickness grows stronger and stronger. Finally, he runs to the wall, climbs up, balances a bit on the top and then jumps to the Eastern side of it. Back in East Germany he is immediately captured by the border police and afterwards interrogated by a person working for the 'Stasi' (short for 'Staatssicherheitsdienst', State Security Service) who points out to him that his reintegration into the GDR will not be unproblematic since he attempted to overcome the border five times so far: 'You are constantly trying to cross the border, either from East to West or from West to East. Where do you really want to be?' Arnulf answers: 'On the other side.' Here Arnulf's existential homelessness becomes obvious for the first time: when he initially arrived in West Germany he associated himself with the West, then Victoria clearly addressed him as part of East Germany, and back in the East he is told that he does not fully belong there either since he once fled the country and several times before attempted to do so. Exploiting his helplessness the 'Stasi' people convince him to work for them which subsequently provides him with the opportunity to move between the two sides (similar to 'aunt Gretchen' in *Die allseitig reduzierte Persönlichkeit – Redupers*).

Subsequently, Arnulf intensifies his relationship with Victoria who at some point coincidentally finds a photo of Andrea in his pocket. Looking at the other woman she realises that the two of them are wearing the same earring (she in her right ear and Andrea in her left one), which in her case is a talisman once given to her by Arnulf. This metaphor can be read as Arnulf splitting a pair of earrings to show his equal love and commitment to two women and the respective side of Germany they represent – a gesture through which he redefines separation as unification: he associates two female strangers (the literal embodiments of East and West Germany) with one another and by doing so anticipates the German reunification on a symbolic and individual level.

Becoming increasingly curious, Victoria finally agrees to visit East Berlin, a decision that Arnulf answers by saying: 'You have accepted it for far too long. You only know half of yourself. The things you are always looking for here but can never find are on the other side.' He thus indicates that a personal reunification is just about to happen with her (the West) meeting the East and – although not intended by Arnulf – Andrea. The latter happens in a café in the East: Arnulf first sees Victoria and steals her passport, which he then slips into Andrea's purse to provide her with the opportunity to leave East Germany. Through this incident the two women meet, get to know each other, go to Andrea's flat together and realise a personality change for only one night: Victoria becomes Andrea and stays in the East and Andrea becomes Victoria, spending the night with Arnulf in the West.

Thus Arnulf comes close to achieving on a personal level what still seems far away on a geopolitical level: the reunification of East and West Germany. However, he believes that he is 'chosen' (in a religious sense) to bring these two 'versions of the same' back together again. Hence he refers to himself as 'Moses' at different stages throughout the film, comparing the German and the Jewish situation and associating the Berlin Wall with the Red Sea. At first we learn about his fascination with the biblical story of Moses and the people of Israel as he reads a book called *The Divided Sea* while being in prison. After his arrival in West Germany he 'preaches' to fellow refugees from Eastern Germany to get involved in his project of working towards the German reunification: 'We are four millions in the West.' And to back up his call for action he talks about the Jews in America (who, according to him, are refugees and live in a diaspora just like him and his Eastern associates): 'Look at the Jews in America! They go for the *whole* thing.' Although he does not manage to get people involved he carries on with his 'reunification mission', driven by visions about the Berlin Wall's sudden opening followed by him leading the reunited German people (in Moses style) to freedom. Moreover, he sprays 'Moses' on the Berlin Wall and founds a pirate radio station, which he calls 'Moses 2, Channel 3 – the station for the four million GDR-emigrates in the FRG'. The latter's first broadcasting is a speech by Arnulf addressing the West German public:

> We are a group of democratically oriented communists and address you for the first time. Our aim is infinite. We hereby declare Germany to be a united and inseparable republic; because it is better to die in the desert than to serve the people of Egypt. However that means: withdrawal of all foreign troops,[44] out of the pacts,[45] tanks into the Amazon, peace on earth. We have to tear down the walls between us, by which I mean to destroy all circumstances within which the individual is a despicable and tortured human being. As long as there is a wall, Andrea, we will never, ... our true face ...

This speech shows Arnulf's desperate attempt to make his voice heard more forcefully and to a wider audience (note his choice of radio as a medium of mass communication or, more importantly, a 'popular geopolitical' means), to gain support ('we are a group'), to bridge the political systems in East and West Germany ('democratically oriented communists'), to get beyond the claustrophobic limitations he experiences in many ways in both parts of Germany ('infinite'), to achieve reunification once and for all ('united and inseparable'), to describe the German experience as diasporic, comparable to the one of the people of Israel (Egypt reference), to get rid of allied control ('foreign troops'), to finally achieve peace ('peace on earth') and thus in effect to free the individual from inhuman living conditions in order to be at ease with oneself and become whole again – making, as in

Der Willi Busch Report, a link between global 'practical geopolitics' and individual existence and/or 'popular geopolitics'.

Still, in the end he has to realise that despite his efforts and tricks there is no way out of the geopolitical dilemma – neither on an individual nor on a national level. Hence he again chooses to climb onto the wall, since it appears to be the border whose diasporic or in-between sense of place he can most identify with (more so than with the ones provided by the Eastern or Western part of Germany). Balancing on the wall he concludes: 'We have to begin to become nobody', a solution which seems to suggest forms of post-nationality and thus moving beyond identifications with particular places since the latter proved to divide and exclude at least as much as they united and included, promoting closure rather than openness.[46]

CONCLUSION: THE UNSETTLING SETTING

The analysis of the above West German 'border narratives' in terms of national identity constructions proved the films to be critical accounts of 'popular geopolitics', thus following a methodological approach which Gilmartin and Kofman, with reference to Sharp, define as 'seek(ing) to identify the geopolitical moments in a variety of representations', in this case film.[47] In opposition to 'traditional' approaches, which are rather narrow in scope (in that they concentrate upon the 'wise men', such as Churchill and Roosevelt), this critical version of 'popular geopolitics' makes space for the everyday whilst still relating the latter to the wider (international or global) power structures as illustrated in this essay. Hence the protagonists' lived experiences were shown as being deeply affected by 'practical geopolitics' (while 'formal geopolitics' remained rather hidden). Willi, Edda and Arnulf have to find ways of coping with the existence of the border – a challenge which all three of them respond to by 'doing popular geopolitics'. As such, they actively incorporate 'the fence', 'the wall', etc. into their day-to-day meaning making practices as a news reporter, an artist/photographer and 'Moses', whereby the border becomes a constituent part of their private and professional lives.

More precisely, Willi, Edda and Arnulf's 'popular geopolitical' actions (individual and collective) show resistance to the 'practical geopolitics' enacted by their state in conjunction with the Western allies – a discourse of separation which they are subjected to but strongly contest. In Routledge's words, they are proponents of a 'geopolitics from below' or an 'anti-geopolitics'.[48] With this term the author describes 'an ethical, political, and cultural force within civil society … that challenges the notion that the interests of the state's political class are identical to the community's interest'.[49] Thus 'anti-geopolitics' can be regarded as a radical form of 'critical geopolitics'. At a later point in the argument, Routledge identifies 'the oppositional discourses of dissident intellectuals'[50] as a form of 'anti-geopolitics' which in the case of the above films refers not only to the protagonists' resistance but to that of the filmmakers who articulate their discontent through their respective productions.

However, this 'resistance from within and behind the scenes' can also be regarded as being born out of a sense of longing for particular 'Heimat' attributes such as wholeness, which means that at a time when 'reunification talk' was largely missing from the official discourse, certain films became forums for articulations of alternative or 'anti-geopolitics'. Hence the border or Berlin Wall, described by Morley and Robins as a 'new national symbol',[51] is still to be seen in relation to the concept of 'Heimat' (the alleged 'old national symbol'). Thus it figures as a programmatic antithesis on its way to synthesis, questioning the possibility of 'Heimat' while at the same time preparing the ground for a new version of it. Therefore, 'border' is not a stable but rather transitionary symbol since people invest in its removal far more than in its maintenance.

In conclusion, West Germany is represented as a site of struggle over meaning. Thus the films perpetuate a sense of place which includes, excludes and combines notions of 'Heimat' and 'border' as well as 'inside', 'no-man's-land' and 'outside'. In this respect, an existential homelessness is probably the sense of place that can be described as the most uniting feature of the films discussed above, rendering places 'no longer the clear supports of our identity'.[52] This does not mean that, in Rose's terms, we move towards a 'non-identification' with place[53] but rather towards its complication. Accordingly, West Germany's post-war sense of place appears to be still considerably influenced by past imaginings, which situates it in a post-"Heimat" position as it has not (yet) produced new "non-Heimat" related narratives of itself.

In conceptual terms this requires an understanding of the relationship between place and identity as an ambiguous one, since positive and negative identifications give rise to both homelessness and the desire for an as yet unrepresentable home. This uncomfortably hybrid form of a sense of place emerges through the interplay between the traditional 'Heimat' imaginary and post-war German reality (e.g. the Berlin Wall), which in many ways can be seen as its antithesis. Thus a sense of 'homelessness in hope for home' is the result of the attempt to not let go of either and to see (contemporary) reality as a transitionary phenomenon, which has to share its identificationary function with seemingly timeless forms of place-based imaginaries. However, these contradictory senses of place seem to a large extent irreconcilable and thus cause conflicting feelings about home which can be maybe best described by Faust's exclamation: 'Zwei Seelen wohnen, ach, in meiner Brust!' (Two souls, ah, inhabit my breast!) The border pre-empts and 'Heimat' prevails – a pairing which causes the individual to be caught between different forms of identification while simultaneously inhabiting them.

NOTES

1. M. Gilmartin and E. Kofman, 'Geopolitics, Empire and Imperialism', in L. Staeheli, E. Kofman and L. Peake (eds), *Feminist Perspectives on Political Geography* (New York: Routledge 2004) pp.1–13.

2. Of course, one could also examine films from the former GDR and USA. However, this would be beyond the scope of the article which will concentrate upon West German representations of itself.

3. G. Ó Tuathail, *Critical Geopolitics* (Minneapolis, MN: University of Minnesota Press 1996).

4. J.P. Sharp, 'Hegemony, Popular Culture and Geopolitics: The *Reader's Digest* and the Construction of Danger', *Political Geography* 15/6–7 (1996) p.557.

5. G. Ó Tuathail, *Rethinking Geopolitics* (London: Routledge 1998) p.3.

6. Ibid.

7. Ibid. pp.4–5.

8. Ibid. pp.5–6.

9. Gilmartin and Kofman (note 1).

10. Ó Tuathail, *Rethinking Geopolitics* (note 5) pp.6–7.

11. T. Elsaesser, *New German Cinema – A History* (London: Macmillan.1989) p.280.

12. C.S. Gray and G. Sloan (eds), *Geopolitics, Geography and Strategy* (London, OR: Frank Cass 1999) p.5.

13. During my research for this essay I came across two articles which chose a geopolitical perspective to analyse German matters in what seems to me a symptomatic way of addressing the question. These are K. Hörschelmann, 'Breaking Ground – Marginality and Resistance in (Post) Unification Germany', *Political Geography* 20/8 (2001) pp.981–1004; W. Natter, 'Geopolitics in Germany, 1919–45', in J. Agnew, K. Mitchell and G. Toal (eds), *A Companion to Political Geography* (Oxford, Blackwell 2003) pp.187–203. Yet there was no publication concerned with either of the two Germanies, the Federal Republic of Germany (FRG) or the German Democratic Republic (GDR), between 1945 and 1989.

14. GoGwilt 1998 p.56.

15. Ibid., p.50.

16. Gilmartin and Kofman (note 1), K. Dodds, *Geopolitics in a Changing World* (Harlow: Pearson Education 2000); G. Ó Tuathail, (Dis)placing Geopolitics: Writing on the Maps of Global Politics', *Society and Space* 12 (1994) pp.525–46.

17. Hereby Dodds largely focuses on production and reception issues, supplying facts and figures to support his argument, yet, only hints at the content of the films he refers to.

18. Ó Tuathail, '(Dis)placing Geopolitics' (note 16) p.540.

19. Ibid. p.541.

20. Steve Bell is the popular satirical cartoonist in the *Guardian* newspaper, commenting on geopolitical issues.

21. Ó Tuathail, *Rethinking Geopolitics* (note 5) p.11.

22. Dodds (note 16) p.84.

23. Elsaesser (note 11); T. Corrigan, *New German Cinema – The Displaced Image* (Indianapolis, IN: Indiana University Press 1983); S. Hake, *National Cinema* (London: Routledge 2002); H.G. Pflaum and H.H. Prinzler, *Film in der Bundesrepublik Deutschland: Der neue deutsche Film von den Anfängen bis zur Gegenwart* (Munich: Hanser 1992); J. Sandford, *The New German Cinema* (London: Oswald Wolff 1980); J. Franklin, *New German Cinema: From Oberhausen to Hamburg* (London: Columbus Books 1983); A. Kaes, *From Hitler to Heimat – The Return of History as Film* (Cambridge, MA: Harvard University Press 1989); K. Phillips, *New German Filmmakers – From Oberhausen through the 1970s* (New York: Frederick Ungar 1984).

24. Elsaesser (note 11) p.216.

25. During the Third Reich 'Heimat' became appropriated by Nazi geopolitics through the concept of 'Lebensraum' which proposes the form of a 'pure' imagined community. For a further discussion of 'Heimat' in relation to film, please see below.

26. Pflaum and Prinzler (note 23).

27. From the two NGC films set in Berlin and concerned with border issues Sander's *Die allseitig reduzierte Persönlichkeit – Redupers* (Redupers, 1977) was chosen in favour of Wenders' *Der Himmel über Berlin* (Wings of Desire, 1987) for two reasons. First, it adds a crucial female perspective to the investigation of the NGC's 'critical geopolitics' and thus enables the latter to explore the gendered nature of geopolitics in NGC productions. Secondly, Wender's famous film has been the object of so many analyses (see, for example, R. Bromley, *From Alice to Buena Vista: The Films of Wim Wenders* (Westport,

CT: Praeger 2001)) that it seemed more appropriate to include a less known, yet in terms of this paper's geopolitics discussion at least equally valuable film.

28. Unless otherwise noted, all translations into English are my own.

29. Ó Tuathail, *Rethinking Geopolitics* (note 5) p.11.

30. Dodds (note 16) p.84. Another, subsidiary aim was to counter the aforementioned theoretical/historical approaches centred around a set group of acclaimed filmmakers and their work through a close analysis of productions (by less known NGC directors) which, since scarcely commented on, provide new insights into geopolitical as well as filmic issues. To substantiate the latter argument, this body of film literature was scrutinised accordingly, and the result corresponds with the thematic absence of 'border' related concerns. Thus in Sandford (note 23), Franklin (note 23), Kaes (note 23), and Phillips' (note 23) works the three films are not mentioned at all. In the film historical accounts of Elsaesser (note 11), Hake (note 23), as well as Pflaum and Prinzler (note 23) the films are briefly referred to as part of the NGC's body of films. Only in Corrigan's book does one find a chapter on Sander's *Die allseitig reduzierte Persönlichkeit – Redupers* (1977). Far from dealing with geopolitical questions, however, the author describes the film as a landmark of women's cinema in West Germany and celebrates it as 'perhaps the most internationally acclaimed feminist film of the New German Cinema'; Corrigan (note 23) p.111. Regarding this, the 'critical geopolitics' of the narrative is sidestepped, which allows for the latter's original study together with N. Schilling, *Der Willi Busch Report* (*The Willi Busch Report*, 1979) and P. Schneider, *Der Mann auf der Mauer* (*The Man on the Wall*, 1982) within the analytical part of this essay.

31. C. Seidl, *Der deutsche Film der fünfziger Jahre* (Munich: Wilhelm Heine Verlag 1987).

32. B. Westermann, *Nationale Identität im Spielfilm der fünfziger Jahre* (Frankfurt am Main: Peter Lang 1990) p.57.

33. See for example E. Boa and R. Palfreyman, *Heimat: A German Dream – Regional Loyalties and National Identity in German Culture 1890–1990* (Oxford: Oxford University Press 2000).

34. D. Morley and K. Robins, 'No Place Like Heimat: Images of Home(land) in European Culture', in E. Carter, J. Donald and J. Squires (eds), *Space and Place: Theories of Identity and Location* (London: Lawrence and Wishart 1993) p.7.

35. E. Santner, *Stranded Objects: Mourning, Memory, and Film in Post-war Germany* (Ithaca, NY: Cornell University Press 1990).

36. Morley and Robbins (note 34) p.17.

37. The italicised headings are not part of the voiceover but structuring and analytical devices inserted by the author.

38. L. Mulvey, 'Visual Pleasure and Narrative Cinema', in C. Penley (ed), *Feminism and Film Theory* (London, BFI 1998) pp.62–3.

39. This theme reappears in *Der Mann auf der Mauer*, where it is broadened however, to incorporate an objectified and feminised view of both East and West Germany, depending upon the relative perspective of the protagonist at the time, as he constantly crosses the border. Further, this theme can be seen in the most famous of German 'border films' *Der Himmel über Berlin* (Wings of Desire, Wenders 1987), where the Berlin wall can be read as a border between the spiritual life of Bruno Ganz's angel and the earthly existence of the woman he silently observes then falls in love with. Here her character can be read as the traditional (or biblical) female paradigm, in which feminine virtues are earthly, physical and thus sexualised. This being in opposition to the masculine and 'spiritual' character of the angel.

40. The title of the film suggests an East–West dialogue: 'Die allseitig reduzierte Persönlichkeit' (the wholly reduced personality), in short Redupers, is a reference and inversion of the GDR's flowery phrase 'Die allseitig verwirklichte sozialistische Persönlichkeit' (the wholly fulfilled socialist personality).

41. This concept, introduced by J. Butler, *Gender Trouble* (London: Routledge 1990), describes the relationship between discourse and the materiality of bodies/things as being indissoluble. Thus the author defines the 'performative' as 'that discursive practice which enacts or produces that which it names' (p.13).

42. Certain groups of people, i.e. pensioners and children, were allowed to move (relatively) freely between East and West Germany since the risk of them not returning was seen as relatively small by officials.

43. In this vein, the 'Willi' version of 'popular geopolitics' seems to still contain certain 'traditional' residues, focusing on the white male, while the 'Edda' version of 'popular geopolitics' appears to be much more overtly 'critical', taking the perspective of a single mother in a patriarchal society. However, even though this is the case to some extent both films, too, offer a critique of solely 'traditional geopolitical' approaches by transgressing the male/active/powerful vs female/passive/powerless binaries. Because of

the latter and by acknowledging and illustrating the gendered nature of geopolitics both narratives can be ultimately associated with the concept of 'critical geopolitics'.

44. This can be read as a call for 'decolonising' Germany, addressing the issue that until today the latter hosts a large number of American and British soldiers in various army bases all over the country.

45. This refers to the NATO (for West Germany) and the Warsaw Pact (for East Germany).

46. Besides his promotion of Germany's 'physical decolonisation' Arnulf here programmatically formulates the idea of 'decolonising the (German) mind' which ultimately means to erase the existing national, 'traditional geopolitical' imagination in order to be able to picture a new German way of being-in-the-world; Ó Tuathail, *Critical Geopolitics* (note 3) p.256.

47. J.P. Sharp, *Condensing Communism: The Reader's Digest and American Identity 1922–1994* (Minneapolis, MN, University of Minneapolis Press 2000); idem, 'Remasculinising Geo-politics? Comments on Gearoid O'Tuathail's *Critical Geopolitics'*, *Political Geography* 19/3 (2000) pp.361–4; in Gilmartin and Kofman (note 1).

48. P. Routledge, 'Introduction to Anti-Geopolitics', in G. Ó Tuathail, S. Dalby and P. Routledge (eds), *The Geopolitics Reader* (London: Routledge 1998) p.245.

49. P. Routledge, 'Anti-Geopolitics', in J. Agnew, K. Mitchell and G. Ó Tuathail (eds), *A Companion to Political Geography* (Oxford: Blackwell 2003) p.236.

50. Ibid. p.237.

51. Morley and Robins (note 34) p.17.

52. Ibid. p.5.

53. Rose, G. 'Place and Identity: A Sense of Place', in D. Massey, P. Jess (eds), *A Place in the World: Places Cultures and Globalization* (Oxford: Oxford University Press 1995) pp.96–7.

Index